Medical Microbiology and Infection at a Glance

Companion website

This book is accompanied by a companion website:

www.ataglanceseries.com/medicalmicrobiology

The website includes:
• Interactive self-assessment case studies
• Summaries of key points for each chapter
• A list of selected online further reading
• Extra images from the authors' digital image bank

Medical Microbiology and Infection at a Glance

Stephen H. Gillespie

MD, DSc, FRCP (Edin), FRCPath
The Sir James Black Professor of Medicine
The Medical School
University of St Andrews
St Andrews
UK

Kathleen B. Bamford

MD, FRCPath
Consultant Medical Microbiologist and Visiting Professor
Imperial College Healthcare NHS Trust and Imperial College London
Hammersmith Hospital
London
UK

Fourth edition

WILEY-BLACKWELL

A John Wiley & Sons, Ltd., Publication

This edition first published 2012, © 2012 by John Wiley & Sons, Ltd.

Previous editions © 2000, 2003, 2007 by Stephen H. Gillespie, Kathleen B. Bamford

Wiley-Blackwell is an imprint of John Wiley & Sons, formed by the merger of Wiley's global Scientific, Technical and Medical business with Blackwell Publishing.

Registered office: John Wiley & Sons, Ltd, The Atrium, Southern Gate, Chichester, West Sussex, PO19 8SQ, UK

Editorial offices: 9600 Garsington Road, Oxford, OX4 2DQ, UK
The Atrium, Southern Gate, Chichester, West Sussex, PO19 8SQ, UK
111 River Street, Hoboken, NJ 07030-5774, USA

For details of our global editorial offices, for customer services and for information about how to apply for permission to reuse the copyright material in this book please see our website at www.wiley.com/wiley-blackwell.

Wiley also publishes its books in a variety of electronic formats. Some content that appears in print may not be available in electronic books.

Library of Congress Cataloging-in-Publication Data
Gillespie, S. H.
  Medical microbiology and infection at a glance / Stephen H. Gillespie, Kathleen B. Bamford. – 4th ed.
      p. ; cm. – (At a glance)
  Includes bibliographical references and index.
  ISBN 978-0-470-65571-9 (pbk. : alk. paper)
  I. Bamford, Kathleen B.   II. Title.   III. Series: At a glance series (Oxford, England).
  [DNLM: 1. Microbiology.   2. Communicable Diseases.   3. Infection. QW 4]
  616.9'041–dc23

                                        2011042658

A catalogue record for this book is available from the British Library.

Set in 9/11.5 pt Times by Toppan Best-set Premedia Limited
Printed and bound in Malaysia by Vivar Printing Sdn Bhd

1   2012

1006670070

Contents

Companion website

This book is accompanied by a companion website:

www.ataglanceseries.com/medicalmicrobiology

The website includes:
• Interactive self-assessment case studies
• Summaries of key points for each chapter
• A list of selected online further reading
• Extra images from the authors' digital image bank

Preface to the fourth edition

The pace of change in infection remains rapid with new infectious agents, treatments and vaccines being discovered, developed and introduced respectively. The emphasis of the subject changes subtly every year, most notably in relation to hospital-acquired infection, with changes in the epidemiology and severity of the pathogens that affect patients both within healthcare facilities and after they have been discharged. The methodologies applied to diagnosis have made rapid progress with an increasing proportion of all infections being detected by molecular means. The availability of high-throughput genome sequencing is influencing our understanding of what makes a pathogen, how it causes disease and how it is transmitted. In this edition we have tried to capture the critically important changes in all these areas so that the information available to the student is as up to date as possible.

Medical Microbiology and Infection at a Glance is intended to provide a distillate of the key facts and principles of infection. We hope that it supports your learning and forms a basis for wider and deeper reading in an important and fascinating subject area of critical importance to all healthcare professionals.

Stephen H. Gillespie
Kathleen B. Bamford

Preface to the first edition

This book is written for medical students and doctors who are seeking a brief summary of microbiology and infectious diseases. It should prove useful to those embarking on a course of study and assist those preparing for professional examinations.

Chapters are divided into concepts, the main human pathogens and the infectious syndromes. This broadly reflects the pattern of teaching in many medical schools.

Microbiology is a rapidly growing and changing subject: new organisms are constantly being identified and our understanding of the pathogenic potential of recognized pathogens is being expanded. In addition the taxonomists keep changing the names of familiar friends to add to the confusion. Despite this, there are clear fundamental facts and principles that form a firm foundation of knowledge on which to build throughout a professional career. It is these that this book strives to encapsulate.

Each chapter contains a diagram which illustrates core knowledge. The associated text offers further insights and details where necessary.

Irrespective of a doctor's specialty, diligent study of microbiology provides the basis for sound professional judgement, giving the clinician confidence and benefiting patients for years to come.

The authors gratefully acknowledge the editorial work of Dr Janet Gillespie who has reminded the authors of practice in a community setting. They are also grateful to Dr Deenan Pillay for his critical reading of the virology sections.

Stephen Gillespie & Kathleen Bamford
London, 2000

1 Structure and classification of bacteria

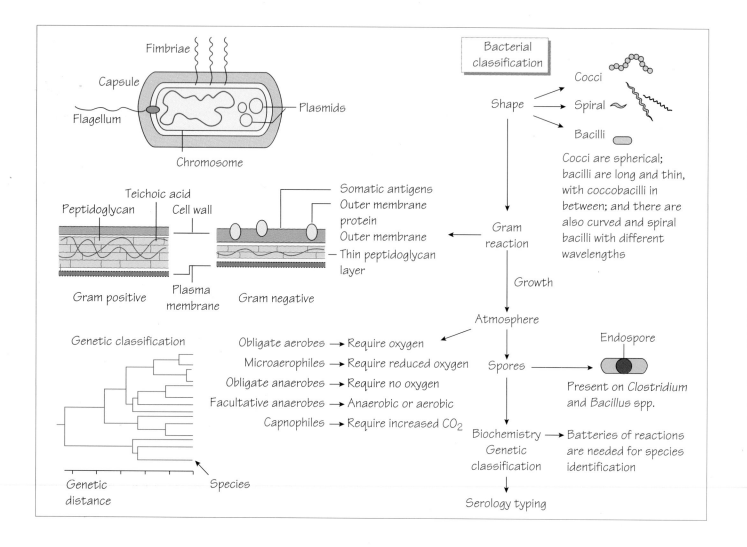

Bacterial classification is important, revealing the **identity** of an organism so that its behaviour and likely response to treatment can be predicted.

Bacterial structural components

Bacterial cell walls are rigid and protect the organism from differences in osmotic tension between the cell and the environment. Gram-positive cell walls have a thick peptidoglycan layer and a cell membrane, whereas Gram-negative cell walls have three layers: inner and outer membranes, and a thinner peptidoglycan layer. The mycobacterial cell wall has a high proportion of lipid, including immunoreactive antigens. Bacterial cell shape can also be used in classification. The following cell components are important for classification, pathogenicity and therapy.

• **Capsule**: a polysaccharide layer that protects the cell from phagocytosis and desiccation.

• **Lipopolysaccharide**: surface antigens that strongly stimulate inflammation and protect Gram-negative bacteria from complement-mediated lysis.

• **Fimbriae** or **pili**: specialized thin projections that aid adhesion to host cells. *Escherichia coli* that cause urinary tract infections bind to mannose receptors on ureteric epithelial cells by their P fimbriae. Fimbrial antigens are often immunogenic but vary between strains so that repeated infections may occur (e.g. *Neisseria gonorrhoeae*).

• **Flagella**: these allow organisms to find sources of nutrition and penetrate host mucus. The number and position of flagella may help identification.

• **Slime**: a polysaccharide material secreted by some bacteria that protects the organism against immune attack and eradication by antibiotics when it is growing in a biofilm in a patient with bronchiectasis or on an inserted medical device.

• **Spores**: metabolically inert bacterial forms adapted for long-term survival in the environment, which are able to regrow under suitable conditions.

Bacteria have a single chromosome and lack a nucleus (prokaryotes). The DNA is coiled and supercoiled by the DNA gyrase enzyme system (see Chapter 6). Bacterial ribosomes differ from

Medical Microbiology and Infection at a Glance, Fourth Edition. Stephen H. Gillespie, Kathleen B. Bamford. © 2012 John Wiley & Sons, Ltd.

eukaryotic ones, making them a target for antibacterial therapy (see Chapter 5). Bacteria also contain accessory DNA in the form of plasmids, integrons, transponsons and bacteriophages. These may transmit antimicrobial resistance (see Chapter 7) and may also code for pathogenicity factors.

Classification of bacteria

We identify microorganisms to predict their pathogenicity: a *Staphylococcus aureus* isolated from blood is more likely to be causing disease than *Staphylococcus epidermidis*. It is important to identify organisms that spread widely in the community and cause serious disease, such as *Neisseria meningitidis*, so that preventative measures can be taken (see Chapter 21). Bacteria are identified using phenotypic, immunological or molecular characteristics.

• **Gram reaction:** Gram-positive and Gram-negative bacteria respond to different antibiotics. Other bacteria (e.g. mycobacteria) may require special staining techniques.
• **Cell shape:** Bacteria may be shaped as cocci, bacilli or spirals.
• **Endospore:** The presence, shape and position of the endospore within the bacterial cell are noted.
• **Fastidiousness:** Certain bacteria have specific O_2/CO_2 requirements, need special media or grow only intracellularly.
• **Key enzymes:** Some bacteria lack certain enzymes, for example, lack of lactose fermentation helps distinguish salmonellae from *E. coli*.
• **Serological reactions:** Interaction of antibodies with surface structures may for example help to distinguish subtypes of salmonellae, *Haemophilus* and meningococcus.
• **DNA sequences:** DNA sequencing of key genes (e.g. 16S ribosomal RNA or DNA gyrase) can identify the organism precisely.

The classification systems are helpful, but strains differ in pathogenicity and virulence within a species and there are similarities across species. For example, some strains of *E. coli* may cause similar diseases to *Shigella sonnei*.

Medically important groups of bacteria

Gram-positive cocci are divided into two main groups: the staphylococci (catalase-positive), the major pathogen being *S. aureus*; and the streptococci (catalase-negative), the major pathogens being *Streptococcus pyogenes*, which causes sore throat and rheumatic fever, and *S. agalactiae*, which causes neonatal meningitis and pneumonia (see Chapters 14 and 15).

Gram-negative cocci include the pathogenic *N. meningitidis*, an important cause of meningitis and septicaemia, and *N. gonorrhoeae*, the agent of urethritis (gonorrhoea).

Gram-negative coccobacilli include the respiratory pathogens *Haemophilus* and *Bordetella* (see Chapters 17 and 22) and zoonotic agents, such as *Brucella* and *Pasteurella* (see Chapter 22).

Gram-positive bacilli are divided into sporing and non-sporing. The sporing types are subdivided into those that are aerobic (*Bacillus*; see Chapter 16) and those that are anaerobic (*Clostridium*; see Chapter 19). Pathogens include *Bacillus anthracis*, which causes anthrax, and clostridia, which cause pseudomembranous colitis, tetanus and, more rarely, gas gangrene and botulism. The non-sporing pathogens include *Listeria* and corynebacteria (see Chapter 16).

Gram-negative bacilli (including the family Enterobacteriaceae) form part of the normal flora of humans and animals, and can be found in the environment. They include many pathogenic genera: *Salmonella*, *Shigella*, *Escherichia*, *Proteus* and *Yersinia* (see Chapter 24). *Pseudomonas* and *Burkholderia* are environmental saprophytes that are naturally resistant to antibiotics and are important hospital pathogens (see Chapter 26). *Legionella* lives in the environment in water but can cause human infection if conditions in the built environment allow it to gain a foothold (see Chapter 26).

Spiral bacteria include the small gastrointestinal pathogen *Helicobacter* that colonizes the stomach, leading to gastric and duodenal ulcers and gastric cancer, and *Campylobacter* spp. that cause acute diarrhoea (see Chapter 25). The *Borrelia* may cause a chronic disease of the skin joints and central nervous system, Lyme disease (*Borrelia burgdorferi*), or rarely relapsing fever (*Borrelia duttoni* and *Borrelia recurrentis*). The *Leptospira* are zoonotic agents that cause an acute meningitis syndrome that may be accompanied by renal failure and hepatitis. The *Treponema* include the causative agent of syphilis (*Treponema pallidum*).

Mycoplasma and Chlamydia are responsible for common respiratory and sexually transmitted infections.

Rickettsia are the agents of typhus and rarer severe infections (see Chapter 27).

2 Innate immunity and normal flora

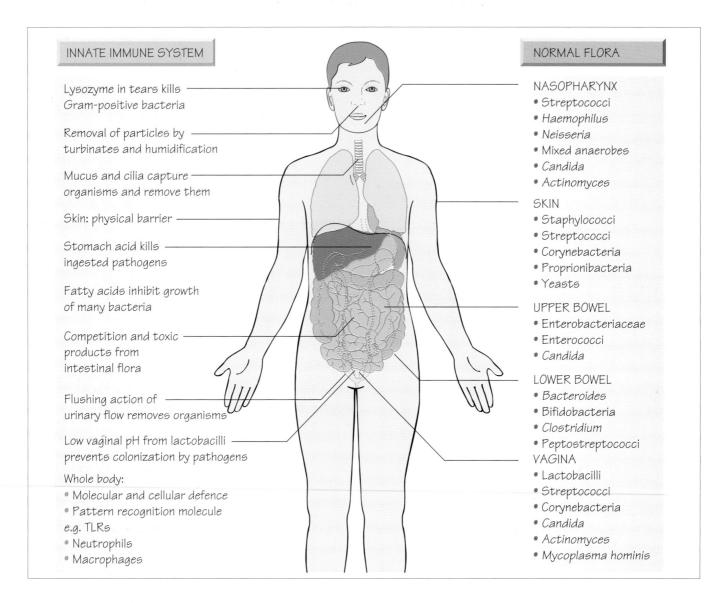

INNATE IMMUNE SYSTEM

Lysozyme in tears kills Gram-positive bacteria

Removal of particles by turbinates and humidification

Mucus and cilia capture organisms and remove them

Skin: physical barrier

Stomach acid kills ingested pathogens

Fatty acids inhibit growth of many bacteria

Competition and toxic products from intestinal flora

Flushing action of urinary flow removes organisms

Low vaginal pH from lactobacilli prevents colonization by pathogens

Whole body:
- Molecular and cellular defence
- Pattern recognition molecule e.g. TLRs
- Neutrophils
- Macrophages

NORMAL FLORA

NASOPHARYNX
- Streptococci
- Haemophilus
- Neisseria
- Mixed anaerobes
- Candida
- Actinomyces

SKIN
- Staphylococci
- Streptococci
- Corynebacteria
- Proprionibacteria
- Yeasts

UPPER BOWEL
- Enterobacteriaceae
- Enterococci
- Candida

LOWER BOWEL
- Bacteroides
- Bifidobacteria
- Clostridium
- Peptostreptococci

VAGINA
- Lactobacilli
- Streptococci
- Corynebacteria
- Candida
- Actinomyces
- Mycoplasma hominis

The innate immune system, which consists of the normal flora, physical barriers such as the skin, antibacterial proteins and phagocytic cells, is an important defence mechanism against infection. Many responses to 'harm' are detected by pattern recognition molecules such as the Toll-like receptors (TLRs), which trigger cascades that activate phagocytes and the immune response. For example, TLR-4 recognizes lipopolysaccharide and TLR-9 recognizes unmethylated CpG dinucleotides. The main components of the system are listed in the Table. Variation in the expression/composition of each component affects an individual's resistance to infection.

Normal flora

Bacterial cells forming part of the normal flora outnumber human cells in the body. The normal flora provides protection by competing with pathogens for colonization sites and producing antibiotic substances (bacteriocins) that suppress other bacteria. Anaerobic bacteria produce toxic metabolic products and free fatty acids that inhibit other organisms. In the female genital tract lactobacilli produce lactic acid that lowers the pH, so preventing colonization by pathogens.

Antibiotics suppress normal flora, which allows colonization and infection by naturally resistant organisms, such as *Candida albicans*. The infective dose of *Salmonella typhi* is lowered by concomitant antibiotic use. Antibiotics may upset the balance between organisms of the normal flora, allowing one to proliferate disproportionately, for example *Clostridium difficile*, which results in a severe diarrhoeal disease (see Chapter 19).

Physical and chemical barriers

The skin provides a physical barrier, with secreted sebum and fatty acids inhibiting bacterial growth. Many pathogens can pen-

Medical Microbiology and Infection at a Glance, Fourth Edition. Stephen H. Gillespie, Kathleen B. Bamford. © 2012 John Wiley & Sons, Ltd.

The innate immune system: the location of barriers to infection, the mechanisms and consequences of deficiency.

Component	Compromise	Consequence
Normal flora		
Pharynx	Antibiotics	Oral thrush
Intestine	Antibiotics	Pseudomembranous colitis; colonization with antibiotic-resistant organisms
Vagina	Antibiotics	Vaginal thrush
Skin	Burns, vectors	Cutaneous bacterial infection, infection with pathogenic viruses, bacteria, protozoa and metazoa
Turbinates and mucociliary clearance	Kartagener's syndrome, cystic fibrosis, bronchiectasis	Chronic bacterial infection
Lysozyme in tears	Sjögren's syndrome	Ocular infection
Urinary flushing	Obstruction	Recurrent urinary infection
Phagocytes, neutrophils, macrophages	Congenital, iatrogenic, infective	Chronic pyogenic infection, increased susceptibility to bacterial infection
Complement	Congenital deficiency	Increased susceptibility to bacterial infection, especially *Neisseria* and *Streptococcus pneumoniae*

etrate the skin, either via the bite of a vector (e.g. *Aedes aegypti* that transmits dengue) or by invasion through intact skin (e.g. *Leptospira* and *Treponema*). Some organisms colonize mucosal surfaces and use this route to gain access to the body (e.g. *Streptococcus pneumoniae*).

If skin integrity is broken by intravenous cannulation or by medical or non-medical injection, blood-borne viruses, such as hepatitis B or the human immunodeficiency virus (HIV), can be transmitted. Diseases of the skin, such as eczema or burns, permit colonization and invasion by pathogens (e.g. *Streptococcus pyogenes*).

Mucociliary clearance mechanisms protect the respiratory tract. Air is humidified and warmed as it is drawn in by passing over the turbinate bones and through the nasal sinuses. Any particles settle on the sticky mucus of the respiratory epithelium and debris is then transported by the cilial 'conveyor belt' to the oropharynx where it is swallowed. As a result only particles with a diameter of less than 5μm reach the alveoli, so the respiratory tract is effectively sterile below the carina.

Secreted antibacterial compounds include mucus, which contains polysaccharides of similar antigenic structure to the underlying mucosal surface; organisms bind to the mucus and are removed. Other antibacterial compounds secreted by the body include lysozyme in tears, which degrades Gram-positive bacterial peptidoglycan; lactoferrin in breast milk, which binds iron and inhibits bacterial growth; lactoperoxidase, a leucocyte enzyme, which produces superoxide radicals that are toxic to microorganisms.

Gastric acid protects from intestinal pathogens – acid suppression increases the risk of intestinal infection.

Urinary flushing protects the urinary tract, with the flushing action of urinary flow keeping the tract sterile, except near the urethral meatus. Obstruction by stones or tumours, benign pros-

tatic hypertrophy or scarring of the urethra or bladder may cause a reduction of urinary flow and stasis, increasing the risks of subsequent bacterial urinary infection.

Phagocytes

Neutrophils and macrophages ingest particles, including bacteria, viruses and fungi. Opsonins (e.g. complement and antibody) may enhance phagocytic ability; for example, *S. pneumoniae* are not phagocytosed unless their capsule is coated with an anticapsular antibody. The action of macrophages in the reticuloendothelial system is essential for resistance to many bacterial and protozoan pathogens, such as *S. pneumoniae* and malaria. Congenital deficiency of neutrophil function leads to chronic pyogenic infections, recurrent chest infections and bronchiectasis. Following splenectomy, patients have defective macrophage function and diminished ability to remove capsulate organisms from the blood.

Complement and other plasma proteins

Complement is a system of plasma proteins that collaborate to resist bacterial infection, which is activated by antigen–antibody binding (the classical pathway) or by direct interaction with bacterial cell wall components (the alternative pathway). The products of both processes attract phagocytes to the site of infection (chemotaxis), activate phagocytes, cause vasodilatation and stimulate phagocytosis of bacteria (opsonization). The final three components of the cascade form a 'membrane attack complex' that can lyse Gram-negative bacteria. Complement deficiencies render patients susceptible to acute pyogenic infections, especially with *Neisseria meningitidis*, *Neisseria gonorrhoeae* and *S. pneumoniae*.

Transferrin captures iron, which limits the amount available to invading microorganisms. Other acute-phase proteins that are directly antibacterial include mannose-binding protein and C-reactive protein (CRP), which binds to bacteria and activates complement.

Pathogenicity and pathogenesis of infectious disease

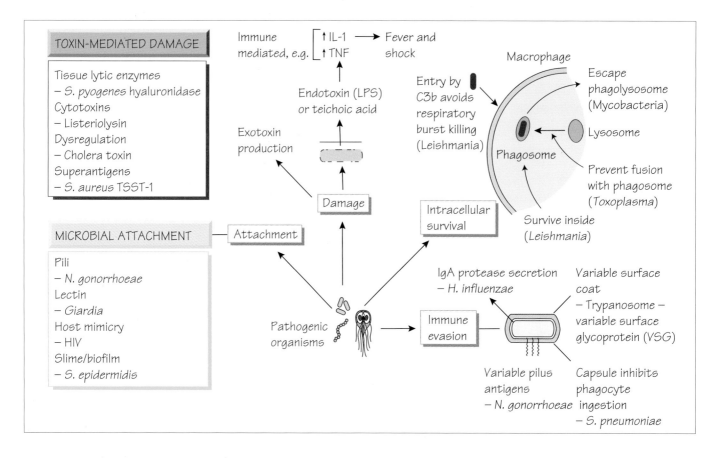

Definitions

Humans encounter bacteria, viruses and parasites that do not cause disease. An infection occurs when microorganisms cause ill-health.

Term	Definition
Pathogen	An organism capable of causing disease
Commensal	An organism that is part of the normal flora
Pathogenicity	The ability to cause disease
Virulence	The ability to cause severe disease

The capsule of *Streptococcus pneumoniae* is a pathogenicity determinant because without it the organism does not usually cause disease. Some capsular types cause more serious disease i.e. they are more virulent (Chapter 15). The term parasite is often used to describe protozoan and metazoan organisms, but this is confusing as these organisms may be either pathogens or commensals.

Types of pathogen

Obligate pathogens are always associated with disease (e.g. *Treponema pallidum* and HIV). Conditional pathogens may cause disease if certain conditions are met. For example, *Bacteroides fragilis* is a normal commensal of the gut but if it invades the peritoneal cavity, it will cause severe infection. Opportunistic pathogens usually cause infection when the host defences are compromised. For example, *Pneumocystis jiroveci* usually causes lung infection only in a host who has severely compromised T-cell immunity.

Mechanisms of pathogenicity

The process of infection has several stages.

Access to a vulnerable host – transmission

Organisms are transmitted by various means but most are restricted to a particular route (Chapter 8). Strains may develop epidemic potential by developing adaptation to an environment that either favours transmission or better survival in that environment. For example, most respiratory pathogens induce coughing, which facilitates their spread by the creation of respiratory droplets. The vomiting and diarrhoea associated with organisms that are spread by the faecal–oral route increase contamination of the environment and the risk of new infection.

Attachment to the host

Microorganisms must attach themselves to host tissues to colonize them and each organism has a different strategy. The distribution

Medical Microbiology and Infection at a Glance, Fourth Edition. Stephen H. Gillespie, Kathleen B. Bamford. © 2012 John Wiley & Sons, Ltd.

of the receptors to which a particular organism can bind will define the organs that it will infect, as in the following examples.

• *Neisseria gonorrhoeae* adheres to the genital mucosa using fimbriae.

• Influenza virus attaches by its haemagglutinin antigen. This accounts for both species-specific pathogenesis (the ability of certain strains to cause disease in a particular species, such as avian or porcine strains) and intraspecies variation in affinity and susceptibility (Chapter 34).

• Bacterial biofilms aid colonization of indwelling prosthetic devices, such as catheters and the respiratory tract. Some strains of staphylococci have genes that mediate attachment to plastics and to the biological molecules that coat intravascular devices.

• Mucosal destruction may expose a variety of host molecules such as fibronectin, vibronectin and collagen to which invading organisms can bind.

Invasion

Some bacteria have mechanisms that help them get close to the mammalian epithelium. For example, *Vibrio cholerae* excretes a mucinase to help it reach the enterocyte. Microorganisms have a variety of strategies that allow them to cross mucosal barriers or different types of cell membrane.

Motility

The ability to move in order to locate new sources of food or in response to chemotactic signals potentially enhances pathogenicity. *V. cholerae* is motile by virtue of its flagellum – non-motile mutants are less virulent.

Immune evasion

To survive in the human host, pathogens must overcome the host immune defences.

• Respiratory bacteria secrete an IgA protease that degrades host immunoglobulin.

• *Staphylococcus aureus* expresses protein A, which binds host immunoglobulin, preventing opsonization and complement activation.

• *S. pneumoniae* has a polysaccharide capsule, which inhibits phagocytosis by polymorphonuclear neutrophils.

• *Toxoplasma gondii*, *Leishmania donovani* and *Mycobacterium tuberculosis* are adapted to survive within macrophages using different mechanisms.

• The lipopolysaccharide (LPS) of Gram-negative organisms makes them resistant to the effect of complement.

• Trypanosoma alter surface antigens to evade antibodies.

Damaging the host
Toxins
Endotoxins

Endotoxins stimulate macrophages to produce cytokines such as interleukin-1 (IL-1) and tumour necrosis factor (TNF) that cause fever and shock.

Exotoxins

Bacterial exotoxins can cause local or distant damage. They are usually proteins and may have a subunit structure.

• Cholera toxin B subunit binds to the epithelial cell and the A subunit activates adenyl cyclase, which results in sodium and chloride efflux from the cell, thus causing diarrhoea.

• Staphylococcal enterotoxins act as superantigens, causing non-specific activation of T cells that have compatible variable region structure, which results in intense cytokine production leading to fever, shock, gastrointestinal disturbance and rash.

• Diphtheria toxin and *Pseudomonas aeruginosa* exotoxin A stop protein production by blocking elongation of proteins.

• Clostridial toxins interfere with neurological or neuromuscular signalling causing, for example, tetanus and botulism.

In many cases antibody to the toxin ameliorates the physiological effects of the disease and is therefore protective (see Chapter 11).

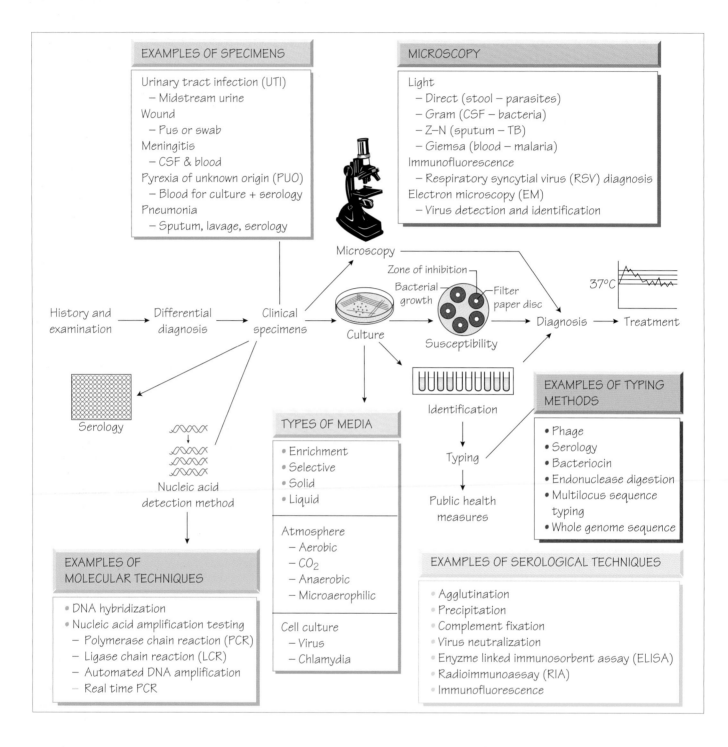

EXAMPLES OF SPECIMENS

Urinary tract infection (UTI)
 – Midstream urine
Wound
 – Pus or swab
Meningitis
 – CSF & blood
Pyrexia of unknown origin (PUO)
 – Blood for culture + serology
Pneumonia
 – Sputum, lavage, serology

MICROSCOPY

Light
 – Direct (stool – parasites)
 – Gram (CSF – bacteria)
 – Z–N (sputum – TB)
 – Giemsa (blood – malaria)
Immunofluorescence
 – Respiratory syncytial virus (RSV) diagnosis
Electron microscopy (EM)
 – Virus detection and identification

Microscopy

Zone of inhibition
Bacterial growth
Filter paper disc

37°C

History and examination → Differential diagnosis → Clinical specimens

Culture → Susceptibility → Diagnosis → Treatment

Serology

Identification

Typing

Public health measures

TYPES OF MEDIA

• Enrichment
• Selective
• Solid
• Liquid

Atmosphere
 – Aerobic
 – CO_2
 – Anaerobic
 – Microaerophilic

Cell culture
 – Virus
 – Chlamydia

Nucleic acid detection method

EXAMPLES OF TYPING METHODS

• Phage
• Serology
• Bacteriocin
• Endonuclease digestion
• Multilocus sequence typing
• Whole genome sequence

EXAMPLES OF MOLECULAR TECHNIQUES

• DNA hybridization
• Nucleic acid amplification testing
 – Polymerase chain reaction (PCR)
 – Ligase chain reaction (LCR)
 – Automated DNA amplification
 – Real time PCR

EXAMPLES OF SEROLOGICAL TECHNIQUES

• Agglutination
• Precipitation
• Complement fixation
• Virus neutralization
• Enyzme linked immunosorbent assay (ELISA)
• Radioimmunoassay (RIA)
• Immunofluorescence

Medical Microbiology and Infection at a Glance, Fourth Edition. Stephen H. Gillespie, Kathleen B. Bamford. © 2012 John Wiley & Sons, Ltd.
Published 2012 by John Wiley & Sons, Ltd.

Specimens

Any tissue or body fluid can be subjected to microbiological investigation with the aim of identifying the infecting pathogen and predicting response to therapy.

To optimize the diagnostic benefit it is necessary to:
• understand in which tissues/specimens the organism is to be found and when in the natural history of the infection;
• take samples carefully (e.g. poor aseptic technique may lead to contamination of sterile samples causing false-positive results);
• transport samples rapidly to the laboratory in a suitable medium.

It is important to remember that some organisms survive poorly outside the body (e.g. numbers of strict anaerobes are reduced by atmospheric oxygen) and in some cases suitable media needs to be inoculated directly in the clinic (e.g. for isolation of *Neisseria gonorrhoeae*).

Laboratory examination

Specimens may be examined grossly, for example to detect adult worms in faeces. Although microscopy is rapid, it is insensitive and requires considerable expertise; specificity may also be a problem if commensal organisms can be mistaken for pathogens. Microscopy can also be used to define specimen quality, for example identifying salivary contamination of sputum by the presence of epithelial cells.

Special stains can be used to help identify organisms, such as Giemsa staining of blood films and tissues, which is used to identify malaria and *Leishmania* (Chapter 42). Immunofluorescence can provide precise identification of a pathogen by using antibodies that are directed against a specific organism.

Culture

• Antibiotic therapy administered presampling can falsely render samples culture negative.
• Most human pathogens are fastidious, requiring media supplemented with nutrients to support growth and increase their numbers to detectable levels.
• Growth on solid media allows organisms to be separated into individual colonies; a pure (clonal) population permits subsequent identification and susceptibility testing.
• Selective agents such as antibiotics or dyes may be used to suppress unwanted organisms in specimens with a normal flora.
• An appropriate atmosphere must be provided: fastidious anaerobes require an oxygen-free atmosphere.
• Most pathogenic bacteria are incubated at 37 °C, but some fungi are incubated at 30 °C.

Identification

• Identification predicts pathogenicity: *Vibrio cholerae* causes severe watery diarrhoea, whereas *Shigella sonnei* infection is usually mild.
• Identification of some organisms should prompt public health action, for example contact tracing for a patient found to have meningococcal meningitis.

Bacterial identification depends on colonial morphology on agar, microscopic morphology, biochemical tests and, increasingly, nucleic acid amplification tests (NAATs) and gene sequencing. This is especially important for organisms that are slow growing (e.g. *Mycobacterium tuberculosis*) or impossible to grow (e.g. *Trophyrema whippelii*) to grow.

Susceptibility testing

Susceptibility testing aims to determine whether treatment with a given antibiotic will be successful. A susceptible organism should respond to a standard dose of an antimicrobial, a moderately resistant strain should respond to a larger dose, whereas a resistant organism is likely to fail therapy with the given antibiotic. Clinical response depends on host factors, and *in vitro* tests only provide an approximate guide to therapy.

Several bodies including the British Society of Antimicrobial Chemotherapy (BSAC) and the Clinical Laboratory Standards Institute (CLSI) define methods and standardized conditions to ensure testing is reproducible. Both are based on measurement of the diameter of the zone of inhibition of confluent growth for the test organism that is caused by an antimicrobial incorporated into a paper disc. The minimum inhibitory concentration, which is the lowest dose that completely inhibits growth, is a more objective method and enables resistance levels to be related to the concentration of antibiotic that is achievable in the tissues.

Susceptibility can be assessed rapidly by hybridization or sequence-based methods that detect specific antibiotic-resistance mutations.

Serology

An infection can be diagnosed by detecting the immune response to the pathogen: for example by detection of rising or falling antibody concentrations more than a week apart, or by the presence of a specific IgM or specific antigen. These techniques are used for organisms that are difficult to grow such as viruses (e.g. HIV or hepatitis B).

Molecular techniques
Southern blotting and nucleic acid hybridization

A labelled DNA probe will bind to the specimen if it contains the specific sequence that is being sought. The captured probe is detected by the activity of it attached label. This technique is specific and rapid, but less sensitive than other methods that involve amplification steps.

Nucleic acid amplification tests

Nucleic acid amplification tests (NAATs) make the diagnosis by amplifying specific regions of the genome from the pathogen. Although different methods are used to amplify pathogen-specific DNA or RNA the aim is the same, to produce sufficient copies for detection. For example, nucleic acid from the pathogen is separated into single strands and primers are designed to bind to target sequences. A polymerase then catalyses synthesis of new nucleic acid and this process is repeated for multiple cycles. Automated systems and commercial kits have made these tests available in many laboratories. Real-time PCR machines measure rising concentrations of target DNA and determine positivity when the concentration passes a set threshold. NAATs have the advantage that they can detect slow-growing organisms or those that are difficult to grow (e.g. *M. tuberculosis*) or make a diagnosis when samples are rendered falsely negative by antibiotic therapy. Methods to detect antibiotic-resistance genes can also be used to provide surrogate susceptibility results (e.g. detection of the *rpoB* gene mutation for rifampicin resistance in *M. tuberculosis*).

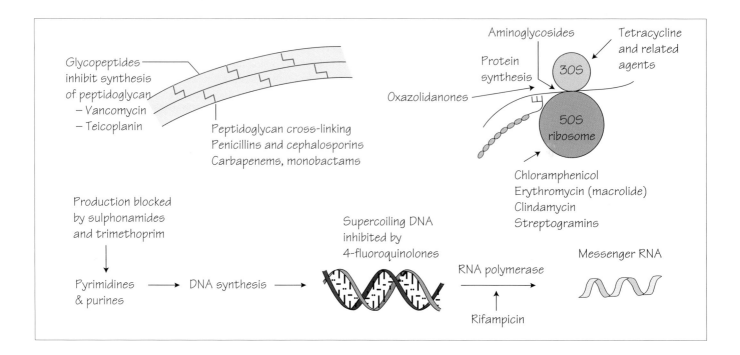

Principles of antibiotic therapy

Antibiotics aim to kill organisms while causing no harm to the patient – this concept is known as **selective toxicity**. It is best achieved by inhibiting bacterial functions that are not present in human cells, for example the peptidoglycan of bacterial cells is inhibited by penicillin. The difference between the dose necessary for treatment and that which causes harm is usually large and is known as the **therapeutic index**. The aminoglycosides are exceptions to this because doses just above the therapeutic level can be toxic. While all antimicrobials have potential unwanted effects, fortunately serious unwanted effects are not frequent.

Choice of therapy

Antibiotic choice depends on the:
- **infection site**: sufficient concentrations of antimicrobial may be difficult to achieve in some sites such as abscesses, bone, the CSF or areas with poor blood supply and some sites with low pH inhibit certain antibiotics (e.g. aminoglycosides);
- **organism**: identification of the organism predicts the natural history of the infection, thereby allowing treatment optimization;
- **susceptibility pattern**: *Streptococcus pyogenes* is invariably susceptible to penicillin, but other organisms such as *Acinetobacter* and *Pseudomonas* are resistant, so antibiotics should be chosen to cover the resistance pattern of all the potential pathogens;
- **severity of infection**: severe infections require antibiotics to be given by the parenteral route;
- **history of allergy**: a previous allergic response may limit the choice of antibiotic;
- **likelihood of unwanted effects**: for example, aminoglycosides should be used with care in patients with pre-existing renal disease.

Routes of administration

Antibiotics are usually taken orally in mild infections, but intravenous therapy is usually required in severe infections, such as septicaemia, to ensure adequate antibiotic concentrations are achieved. Intravenous therapy may also be chosen for patients unable to tolerate oral therapy. Topical administration is suitable for skin infections. More rarely, antibiotics are given *per rectum* (e.g. metronidazole for surgical prophylaxis) or intravaginally as pessaries. Children require palatable liquid formulations to maximize adherence.

Monitoring therapy

It is possible to measure the concentration of an antibiotic in the serum or at the site of infection (e.g. in the CSF for meningitis). This may be necessary to:
- ensure that there are adequate therapeutic concentrations at the site of infection;
- reduce the risk of toxicity, which is important where the therapeutic index is low (serum levels of aminoglycosides should be measured in serum taken just before and 1 h after intravenous or intramuscular dosage, which allows the dose to be adjusted according to normograms and careful adherence to guidelines, e.g. for a high peak the dosage may be reduced or for a high trough level the medication is given less frequently);
- assist in the management of an infection with intermediate susceptibility (if inhibition of an organism occurs only at high antibiotic concentrations, it is important to ensure sufficient concentrations are found at the site of infection, e.g. in *Pseudomonas* meningitis, antibiotic concentrations should be measured in the CSF);

Medical Microbiology and Infection at a Glance, Fourth Edition. Stephen H. Gillespie, Kathleen B. Bamford. © 2012 John Wiley & Sons, Ltd.

• study the pharmacokinetics of the drug (treatment plans are based on knowing the absorption, distribution and protein binding of drugs). In the development of new antibiotics the way in which the new agent is absorbed and distributed throughout the body is studied by careful sampling.

Adverse events
Mild gastrointestinal upset is probably the most frequent side effect of antibiotic therapy. Rarely, severe allergic reactions may lead to acute anaphylactic shock or serum sickness syndromes.

Gastrointestinal tract
Antibiotic activity can upset the balance of the normal flora within the gut (β-lactams are especially likely to do this) resulting in overgrowth of commensal organisms such as *Candida* spp. Alternatively, antibiotic therapy may provoke diarrhoea or, more seriously, pseudomembranous colitis (see Chapter 19).

Skin
Cutaneous manifestations range from mild urticaria or maculopapular, erythematous eruptions to erythema multiforme and the life-threatening Stevens–Johnson syndrome. Most cutaneous reactions are mild and resolve after discontinuation of therapy.

Haemopoietic system
Patients receiving chloramphenicol or antifolate antibiotics may exhibit dose-dependent bone marrow suppression. More seriously, aplastic anaemia may rarely complicate chloramphenicol therapy. High doses of β-lactam antibiotics may induce granulocytopenia. Antibiotics are a rare cause of haemolytic anaemia. Many antibiotics cause a mild reversible thrombocytopenia or bone marrow depression.

Renal system
Aminoglycosides may cause renal toxicity by damaging the cells of the proximal convoluted tubule. Patients who are elderly, have pre-existing renal disease or are also receiving other drugs with renal toxicity are at higher risk. Tetracyclines may also be toxic to the kidneys.

Liver
Isoniazid and rifampicin may cause hepatitis; this is more common in patients with pre-existing liver disease. Other agents associated with hepatitis are tetracycline, erythromycin, pyrazinamide, ethionamide and, very rarely, ampicillin or fluoroquinolones. Cholestatic jaundice may follow tetracycline or high-dose fusidic acid therapy.

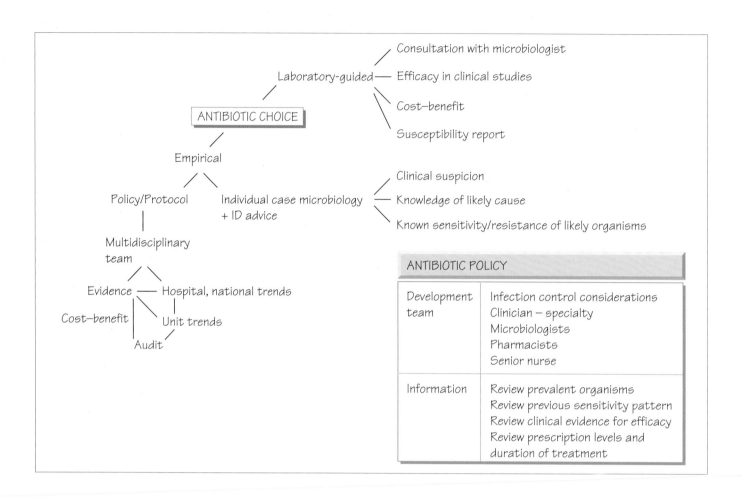

Beta-lactam antibiotics

Penicillins work by inhibiting peptidoglycan cross-linkage. Modifications to the penicillins have extended their antibacterial spectrum and improved absorption. Penicillins now include:

- natural penicillins (e.g. benzylpenicillin, penicillin V);
- penicillinase-resistant penicillin (e.g. flucloxacillin);
- aminopenicillins (e.g. ampicillin-like agents);
- expanded-spectrum penicillins (e.g. piperacillin);
- penicillins combined with β-lactamase inhibitors (e.g. amoxicillin and clavulanate, known as co-amoxiclav).

Oral absorption varies: benzylpenicillin (penicillin G) is unstable in the presence of gastric acid and must be given intravenously, but penicillin V is stable and can be given orally. The aminopenicillins and flucloxacillin are also absorbed orally, while the remaining agents must be given intravenously.

Penicillins are secreted by the kidney and have a short half-life. They are distributed in extracellular fluid, but do not cross the blood–brain barrier unless the meninges are inflamed.

Cephalosporins

Cephalosporins are closely related to penicillins. They are all active against Gram-positive organisms and later compounds have activity against Gram-negative bacteria including *Pseudomonas*.

Monobactams

The monobactams are related to penicillins and cephalosporins. They have a broad spectrum of activity, including against anaerobes. Imipenem and meropenem have antipseudomonal effects. They must be given intravenously.

Aminoglycosides

Aminoglycosides act by preventing translation of mRNA into proteins. They are given parenterally, are limited to the extracellular fluid and are excreted in the urine. Aminoglycosides are toxic to the kidney and eighth cranial nerve at amounts close to therapeutic levels, which necessitates careful monitoring of serum concentrations.

Glycopeptides

The glycopeptides (vancomycin and teicoplanin) have the following characteristics.

- They inhibit peptidoglycan cross-linking in Gram-positive organisms only.
- Resistance to them is rare but sometimes found in enterococci (glycopeptide-resistant enterococci – GRE) and in some *Staphylococcus aureus*.

Medical Microbiology and Infection at a Glance, Fourth Edition. Stephen H. Gillespie, Kathleen B. Bamford. © 2012 John Wiley & Sons, Ltd.

- Administration is intravenous or intraperitoneal; they are not absorbed orally. The exception is the oral use of vancomycin to treat pseudomembranous colitis.
- They are distributed in the extracellular fluid, but do not cross the blood–brain barrier unless there is meningeal inflammation.
- Excretion is via the kidney.

Daptomycin

Daptomycin, a new agent with a long half-life, is very active against Gram-positive organisms demonstrating more rapid killing *in vitro*. Its mode of action is uncertain.

Quinolones

- Quinolones act by inhibiting bacterial DNA gyrase.
- The early quinolones did not attain high tissue levels and were used only for urinary tract infections.
- Fluorine modification (fluoroquinolones) has made them active against Gram-negative pathogens including *Chlamydia*.
- Ciprofloxacin has activity against *Pseudomonas spp.*
- Quinolones are well absorbed orally, are widely distributed and penetrate cells well.
- Newer agents (e.g. moxifloxacin) are active against Gram-positive pathogens, including *Streptococcus pneumoniae* and *Mycobacterium tuberculosis*.

Macrolides

The macrolides (erythromycin, azithromycin and clarithromycin) bind to the 50S ribosome, interfering with protein synthesis; they are active against Gram-positive cocci, many anaerobes (but not *Bacteroides*), *Mycoplasma* and *Chlamydia*. They are absorbed orally, distributed in the total body water, cross the placenta, are concentrated in macrophages, polymorphs and the liver and are excreted in the bile. Erythromycin may cause nausea. The newer macrolides (e.g. azithromycin) have more favourable pharmacokinetic and toxicity profiles.

Streptogramins

Pristinamycin is a bactericidal semisynthetic streptogramin consisting of quinupristin and dalfopristin. It acts by preventing peptide bond formation, which results in release of incomplete polypeptide chains from the donor site. It is active against a broad range of Gram-positive pathogens and some Gram-negatives, such as *Moraxella*, *Legionella*, *Neisseria meningitidis* and *Mycoplasma*.

It is used mainly for the treatment of resistant Gram-positive infections (e.g. GRE and glycopeptide-intermediate *S. aureus* [GISA]).

Oxazolidinones

The oxazolidinones (e.g. linezolid) inhibit protein synthesis at the 50S ribosomal subunit. They are most active against Gram-positive bacteria and are used mainly for the treatment of resistant Gram-positive infections. Linezolid is well absorbed orally and concentrated in the skin.

Metronidazole

The main features of metronidazole are that it is:
- active against all anaerobic organisms;
- a receiver of electrons under anaerobic conditions, so forms toxic metabolites that damage bacterial DNA;
- also active against some species of protozoa, including *Giardia, Entamoeba histolytica* and *Trichomonas vaginalis;*
- absorbed orally and can be administered parenterally;
- widely distributed in the tissues, crossing the blood–brain barrier and penetrating into abscesses;
- metabolized in the liver and excreted in the urine;
- well tolerated, except that it cannot be taken with alcohol.

Tetracyclines

- Tetracyclines act by inhibition of protein synthesis by locking tRNA to the septal site of mRNA.
- They are active against many Gram-positive and some Gram-negative pathogens, *Chlamydia*, *Mycoplasma*, *Rickettsia* and treponemes, *Plasmodium* and *Entamoeba histolytica*.
- Doxycycline is absorbed orally, has a long half-life and is widely distributed; adequate therapeutic levels may be obtained by a once-daily dosage.
- The newer tetracyclines such as tigecycline are used to treat multiresistant Gram-negative infections.

Sulphonamides and trimethoprim

Sulphonamides and trimethoprim act by inhibiting the synthesis of tetrahydrofolate. They are now rarely used in the treatment of bacterial infections but have an important role in the management of *Pneumocystis jiroveci* and protozoan infections including malaria. Sulphonamides can be given intravenously and are well absorbed when given orally. They are widely distributed in the tissues and cross the blood–brain barrier. They are metabolized in the liver and excreted via the kidney.

7 Resistance to antibacterial agents

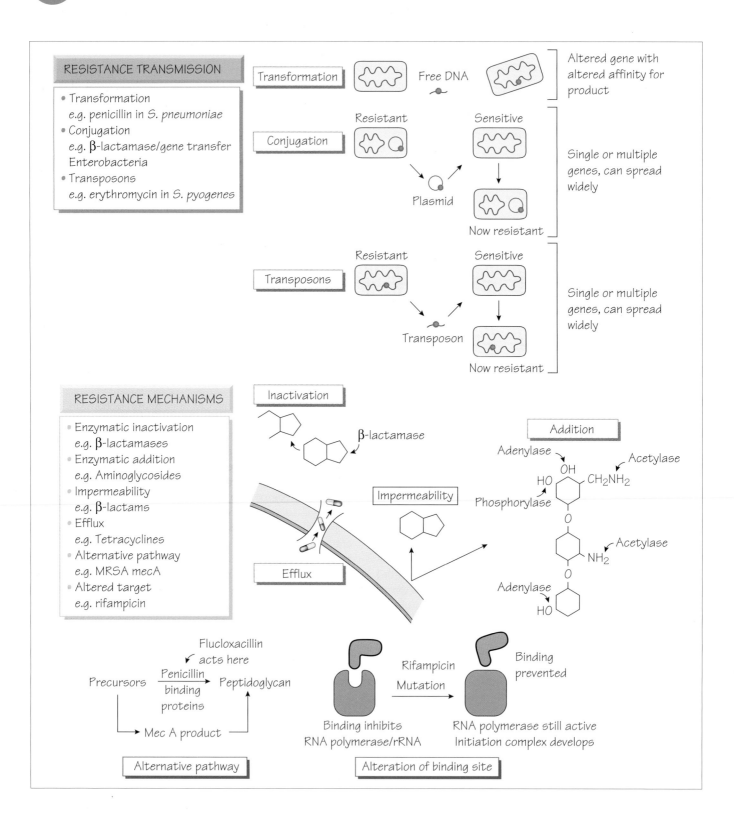

Resistance occurs when a previously susceptible organism is no longer inhibited by an antibiotic at a concentration that can be safely achieved in clinical practice. Resistance can develop quickly because:

- bacteria multiply rapidly;
- mutations arise regularly;
- segments of DNA can transfer by transformation;

Medical Microbiology and Infection at a Glance, Fourth Edition. Stephen H. Gillespie, Kathleen B. Bamford. © 2012 John Wiley & Sons, Ltd.

- genes can be transferred rapidly by bacteriophages, plasmids or other mobile genetic elements.

Antibiotic use favours the survival of resistant organisms. The replication of organisms that have accidentally developed mechanisms to avoid destruction can pose a threat to the successful treatment of infections.

Transmission of resistance determinants between bacteria

Transformation
Many bacterial species incorporate naked DNA into their genome, a process called transformation. For example, *Streptococcus pneumoniae* and *Neisseria gonorrhoeae* incorporate small sections of penicillin-binding protein genes from closely related species to produce a penicillin-binding protein that binds penicillin less avidly, so becoming more resistant. Such organisms are still able to synthesize peptidoglycan and maintain their cell walls in the presence of penicillin.

Conjugation
Bacteria contain plasmids, circular DNA structures that are found in the cytoplasm. Many genes are carried on plasmids including those that encode metabolic enzymes, virulence determinants and antibiotic resistance. Plasmids can pass from one bacterium to another by conjugation allowing 'resistance genes' to spread rapidly in populations of bacterial species that share the same environment (e.g. within the intestine). Combined with antibiotic selective pressures (e.g. in hospitals) that favour the survival of organisms with resistance plasmids, multiresistant populations may develop.

Transposons and integrons
Transposons and integrons are mobile genetic elements able to encode transposition and move between the chromosome and plasmids, and between bacteria. Many functions, including antibiotic resistance, can be encoded on a transposon. Resistance to methicillin among *Staphylococcus aureus* and to tetracycline among *N. gonorrhoeae* probably entered the species by this route. Integrons are important in transmission of multiple drug resistance in Gram-negative pathogens. Resistance genes can also be mobilized by bacteriophages (viruses that live in bacteria).

Multiple resistance
Multiple resistance can develop on mobile genetic elements because once a gene is established on the element, it can readily acquire resistance to another agent by one of the mechanisms above. Once there is more than one resistance gene, exposure to any of these agents will permit survival of the organism, which increases the risk of further resistance being acquired.

Mechanisms of resistance

Antibiotic modification
Enzyme inactivation
A common resistance mechanism is degradation of the antibiotic. Many strains of *S. aureus* produce an extracellular enzyme (β-lactamase), which can break open the penicillin β-lactam ring, thereby inactivating it. In the face of newer β-lactam antibiotics, many human pathogens have acquired a range of genes that encode broad-spectrum β-lactamases; these include *Escherichia coli*, *Haemophilus influenzae* and *Pseudomonas* spp. The genes are often found on mobile genetic elements (transposons). The spread of different types of extended-spectrum β-lactamases (ESBLs), such as CTX-M and AmpC, among Enterobacteriaceae is producing resistance to cephalosporins and broad-spectrum penicillins in organisms that cause hospital-associated infections. Spread to the community has already occurred.

Enzyme addition
Some bacteria express enzymes that add an inactivating chemical group to the antibiotic, so inhibiting its activity. Bacteria may become resistant to aminoglycosides by adding an acetyl, amino or adenosine group to the antibiotic molecule. Different aminoglycosides differ in their susceptibility to this modification, amikacin being the least susceptible. Aminoglycoside-resistance enzymes are found in Gram-positive organisms (e.g. *S. aureus*) and Gram-negative organisms (e.g. *Pseudomonas* spp).

Impermeability
Some bacteria are naturally resistant to antibiotics because their cell envelope is impermeable to that particular antibiotic (e.g. *Pseudomonas* spp. are impermeable to some β-lactam antibiotics). Aminoglycosides enter bacteria by an oxygen-dependent transport mechanism and so have little effect against anaerobic organisms. Other bacteria may lose a porin protein, so creating a permeability barrier that stops antibiotics from entering the cell.

Efflux mechanisms
Bacteria, for example *E. coli* or streptococci, may become resistant to tetracyclines, macrolides or fluoroquinolones by the acquisition of an inner membrane protein that actively pumps the antibiotic out of the cell – an **efflux pump**.

Alternative pathway
Bacteria may acquire genes that create an alternative pathway that can circumvent the metabolic block imposed by an antibiotic. *S. aureus* becomes resistant to methicillin or flucloxacillin when it acquires the gene *mecA*, which encodes an alternative penicillin-binding protein (PBP2′) that is not inhibited by methicillin. Although the composition of its cell wall is altered, the organism is still able to multiply.

Alteration of the target site
Rifampicin acts by inhibiting the β-subunit of RNA polymerase. Resistance develops when the RNA polymerase gene is altered by point mutations, insertions or deletions. The new RNA polymerase is not as easily inhibited by rifampicin and resistance occurs. Similarly, an alteration of the binding sites on DNA gyrase (the target of fluoroquinolones) can make an organism resistant. The genes responsible for these effects are often found in a small region of the target gene, for example in the rifampicin resistance-determining region (RRDR).

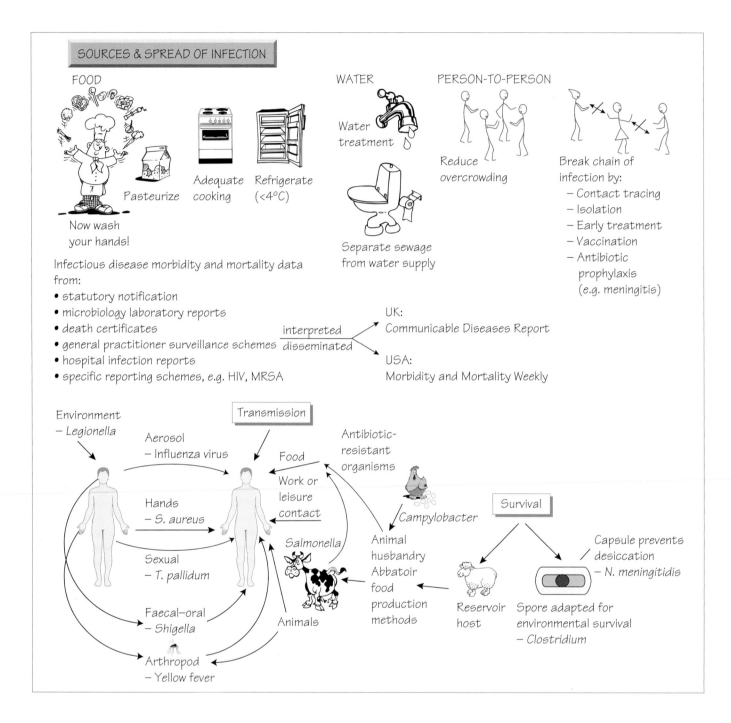

SOURCES & SPREAD OF INFECTION

FOOD

Now wash your hands!

Pasteurize

Adequate cooking

Refrigerate (<4°C)

WATER

Water treatment

Separate sewage from water supply

PERSON-TO-PERSON

Reduce overcrowding

Break chain of infection by:
– Contact tracing
– Isolation
– Early treatment
– Vaccination
– Antibiotic prophylaxis (e.g. meningitis)

Infectious disease morbidity and mortality data from:
- statutory notification
- microbiology laboratory reports
- death certificates
- general practitioner surveillance schemes
- hospital infection reports
- specific reporting schemes, e.g. HIV, MRSA

interpreted disseminated

UK:
Communicable Diseases Report

USA:
Morbidity and Mortality Weekly

Transmission

Environment – Legionella

Aerosol – Influenza virus

Hands – S. aureus

Sexual – T. pallidum

Faecal–oral – Shigella

Arthropod – Yellow fever

Food

Work or leisure contact

Salmonella

Animals

Antibiotic-resistant organisms

Campylobacter

Animal husbandry Abbatoir food production methods

Survival

Reservoir host

Spore adapted for environmental survival – Clostridium

Capsule prevents desiccation – N. meningitidis

Medical Microbiology and Infection at a Glance, Fourth Edition. Stephen H. Gillespie, Kathleen B. Bamford. © 2012 John Wiley & Sons, Ltd.
Published 2012 by John Wiley & Sons, Ltd.

Sources of infection

Infection is caused either by organisms from the host's normal flora (endogenous infection) or by organisms transmitted from another source (exogenous infection).

Endogenous infection

The normal flora will only invade if circumstances permit, as in some of the following examples.
• Following bowel perforation Enterobacteriaceae and non-sporing anaerobes such as *Bacteroides fragilis* can invade the peritoneal cavity causing peritonitis and septicaemia.
• Inhalation of stomach contents may cause pneumonia, possibly followed by a lung abscess.
• *Staphylococcus aureus* that are normally found in the nose may cause wound infection if inoculated into a surgical wound.
• Neutropenic patients are especially prone to infection from organisms normally held in check by the body's defences.
• Surgery and intravenous cannulation allow skin organisms to invade.

Exogenous infection

The most important source of human infection is other humans. Some agents (e.g. measles) are more transmissible than others. An epidemic or outbreak can occur if each infected individual typically transmits the pathogen to more than one additional person.

Animal pathogens may be spread to humans by direct contact or in food. Such infections are called **zoonoses** and are encouraged by intensive farming methods that permit organisms hazardous to humans to spread within the herd or flock (e.g. bovine tuberculosis or *E. coli* O157). The risks increase with increasing intensity of farming: battery hens are especially prone to the spread of *Salmonella* and mechanized meat recovery techniques may increase the likelihood of cross-contamination. Feeding ruminant offal to cattle resulted in an epidemic of bovine spongiform encephalopathy (BSE), which then spread to humans as variant Creutzfeldt–Jakob disease (vCJD). The introduction of good farming and food factory techniques will reduce the risks of zoonoses.

Humans can become infected from organisms in the inanimate environment. For example, poorly maintained air-conditioning cooling towers can be a source of *Legionella pneumophila*.

Survival and transmission

Microorganisms have evolved life cycles that facilitate their transmission and survival.

Organisms that cause diarrhoea, which are excreted in faeces, spread to other hosts by ingestion (faecal–oral route). More complex examples include organisms with a life-cycle stage inside an insect vector that allows transmission by a biting insect.

Organisms have developed specialist structures and behaviours to favour survival.
• Bacterial spores have a tough coat and low metabolic rate that enable them to survive for many years.
• Helminth eggs have a tough shell adapted for survival in the environment.
• The host can aid survival: when the infecting organism is able to persist for a long time in the host, this acts as a **reservoir of infection**.

Airborne/respiratory

Microorganisms propelled from the nose and mouth in a sneeze can remain suspended in the air on droplet nuclei (5 µm). Infection may occur when these are inhaled by another person and are carried to the alveoli. Respiratory infections such as influenza are transmitted in this way, but so are certain infections of other organs (e.g. *Neisseria meningitidis*).

Faecal–oral

Food and water contain pathogens that may infect the intestinal tract (e.g. *Salmonella*). Toxoplasmosis and cysticercosis, which infect organs remote from the gut, are also transmitted by this route.

Parenteral/transcutaneous

• *Leptospira*, *Treponema* and *Schistosoma* have evolved specific mechanisms enabling them to invade intact skin.
• Injections, whether medical or for taking illegal drugs, and blood transfusions bypass the skin, allowing the transmission of a range organisms including the blood-borne viruses HIV and hepatitis B.
• Skin organisms (e.g. *Staphylococcus epidermidis*) can invade the body via indwelling venous cannulae.

Vector-borne

Insects that feed on blood may transmit a wide range of pathogens: most importantly female anophelene mosquitoes transmit malaria.

Sexual transmission

Sexual intercourse allows organisms with poor survival ability outside the body to be transmitted. Examples include *Neisseria gonorrhoeae*, *Treponema pallidum* and HIV. Transmission of HIV is enhanced by genital ulceration.

9 Principles of infection control

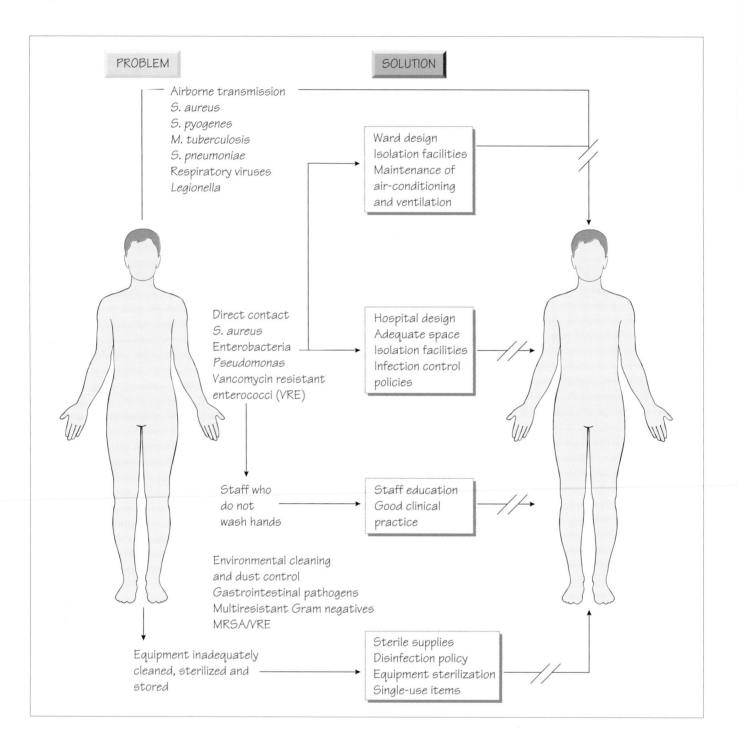

PROBLEM

SOLUTION

Airborne transmission
S. aureus
S. pyogenes
M. tuberculosis
S. pneumoniae
Respiratory viruses
Legionella

Ward design
Isolation facilities
Maintenance of
air-conditioning
and ventilation

Direct contact
S. aureus
Enterobacteria
Pseudomonas
Vancomycin resistant
enterococci (VRE)

Hospital design
Adequate space
Isolation facilities
Infection control
policies

Staff who
do not
wash hands

Staff education
Good clinical
practice

Environmental cleaning
and dust control
Gastrointestinal pathogens
Multiresistant Gram negatives
MRSA/VRE

Equipment inadequately
cleaned, sterilized and
stored

Sterile supplies
Disinfection policy
Equipment sterilization
Single-use items

Medical Microbiology and Infection at a Glance, Fourth Edition. Stephen H. Gillespie, Kathleen B. Bamford. © 2012 John Wiley & Sons, Ltd.
Published 2012 by John Wiley & Sons, Ltd.

Identifying an outbreak

An outbreak, whether in a hospital or the community, must first be recognized by clinical awareness and effective laboratory diagnosis; data must be managed centrally, a process usually described as **surveillance**.

Screening and diagnostics

It is necessary to make a laboratory diagnosis on patients with infection (e.g. to identify methicillin-resistant *Staphylococcus aureus* [MRSA] in the wounds of hospital patients, or influenza in patients in the community). The diagnostic methods, which are described in Chapter 4, should be optimized to identify the organisms of concern. Precise diagnosis often requires the sample or organism to be sent to a reference laboratory. Identification of the organism should be sufficient to permit increased numbers of key pathogens to be recognized. For example, samples/organisms from patients with meningococcal disease should be confirmed and typed at the reference laboratory so that an increased incidence of infection with a particular strain can be detected.

Surveillance systems

Hospitals and communities require a mechanism to collate data as it emerges from laboratories, notifications and other systems, thus allowing an outbreak to be detected. This activity is performed locally by the hospital epidemiologist and nationally by epidemiologists in the public health service, which receives data from the reporting systems at reference laboratories and from other professionals such as environmental health officers.

Mandatory reporting and improvement targets with key measures have focused attention on healthcare-associated infections (e.g. MRSA and *Clostridium difficile*) and have proved beneficial in driving down infection rates.

Typing

Specific testing is used to identify the relationships between organisms: the more closely related organisms are, the more likely they are to have been transmitted from one person to another. Laboratory techniques should be simple to perform, reproducible and give identical results in all laboratories.

Laboratory testing can be divided into simple phenotypic testing and more complex molecular tests. Older methods such as serological and phage typing are being made obsolete by molecular methods.

Molecular typing involves the use of enzymes that can digest genomic, plasmid DNA or ribosomal RNA giving a pattern that can be compared between isolates. Identical organisms will have identical band patterns. Sequence-based methods are emerging, such as multilocus sequence typing (MLST) where a portion of seven housekeeping genes are sequenced to produce a numerical code for the organism. With reducing costs, sequencing of the whole genome may be used where variation is low and the importance of accuracy is high (e.g. a deliberate release of anthrax).

Infection control

Infection can be controlled by separating infected patients from the uninfected before transmission occurs. Methods of achieving this will be detailed in the **infection control policy**. To be successful, this policy must be supported by the entire hospital staff. The control of infection team, which consists of a consultant infection specialist, management representation and specialist nurses, promotes the policy. The infection control committee must be linked to the senior hospital management committee. The team organizes hospital surveillance of high-risk organisms (e.g. MRSA) and must be involved in service development, including for example building alterations and new clinical services.

Isolation

Infected patients should be isolated (source isolation) using methods that will interrupt the means of transmission. Patients who are especially susceptible to infection require protective isolation.

Wound and enteric isolation

For organisms spread by contact or by faecal–oral route, the following precautions should be taken:
• side-room isolation with dedicated hand-washing and toilet facilities;
• use of disposable plastic aprons and gloves for touching the patient;
• discarding of apron and gloves after use, then hand washing with liquid soap and disposable towels.

Disinfection may reduce environmental contamination e.g. chlorine-containing agents for *Clostridium difficile*; see Chapter 19).

Respiratory isolation

For organisms spread by the respiratory route, the following precautions should be taken:
• side-room isolation, as above;
• wearing of facemasks by staff/visitors when in the room;
• wearing of a facemask by the patient if transferred out of their room.

Infection such as multidrug-resistant tuberculosis (MDRTB) and novel respiratory viruses such as pandemic 'flu require negative-pressure facilities and high-efficiency masks or personal respirators.

Strict isolation

This is used for patients with infections that carry a high mortality, such as viral haemorrhagic fevers (see Chapter 38). An enclosed isolation unit with negative-pressure and enclosed-air system, together with strict decontamination procedures, is used to prevent aerosol transmission of the organism.

Protective isolation

Protective isolation is required for patients who are highly susceptible to infection, such those who are neutropenic as a result of radiotherapy or chemotherapy. The following procedures should be followed:
• side-room isolation;
• use of gloves and masks, as in wound and enteric infections;
• provision of filtered air;
• measures to control the risk from organisms in food (e.g. resistant Gram-negative organisms in vegetables or *Listeria* in soft cheeses).

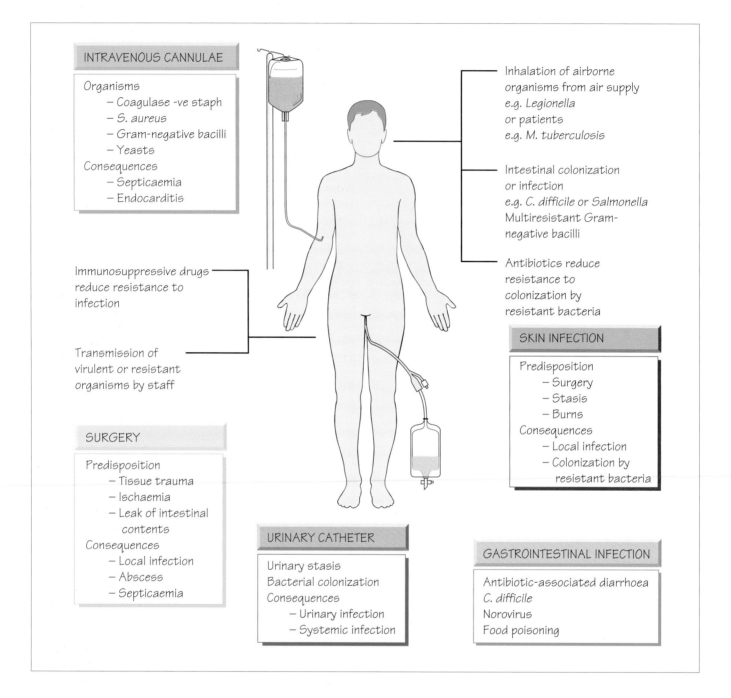

INTRAVENOUS CANNULAE

Organisms
- Coagulase -ve staph
- S. aureus
- Gram-negative bacilli
- Yeasts

Consequences
- Septicaemia
- Endocarditis

Inhalation of airborne organisms from air supply e.g. Legionella or patients e.g. M. tuberculosis

Intestinal colonization or infection e.g. C. difficile or Salmonella Multiresistant Gram-negative bacilli

Antibiotics reduce resistance to colonization by resistant bacteria

Immunosuppressive drugs reduce resistance to infection

Transmission of virulent or resistant organisms by staff

SKIN INFECTION

Predisposition
- Surgery
- Stasis
- Burns

Consequences
- Local infection
- Colonization by resistant bacteria

SURGERY

Predisposition
- Tissue trauma
- Ischaemia
- Leak of intestinal contents

Consequences
- Local infection
- Abscess
- Septicaemia

URINARY CATHETER

Urinary stasis
Bacterial colonization
Consequences
- Urinary infection
- Systemic infection

GASTROINTESTINAL INFECTION

Antibiotic-associated diarrhoea
C. difficile
Norovirus
Food poisoning

Infection is a major cause of mortality and morbidity in patients admitted to hospital. The most frequent types of infection are urinary tract, respiratory, wound, skin and soft-tissue infections, and septicaemia, which is often associated with vascular access.

The environment

Food supply

Food is usually prepared centrally in the hospital kitchens. Patients are at risk of food-borne infection if hygiene standards fall; this route can transmit antibiotic-resistant organisms to immunocompromised patients who are especially vulnerable.

Air supply

Pathogens (e.g. multidrug-resistant tuberculosis, respiratory viruses or bacteria) may be transmitted via theatre air supply and air-conditioning systems. Badly maintained air-conditioning systems may be a source of *Legionella*.

Medical Microbiology and Infection at a Glance, Fourth Edition. Stephen H. Gillespie, Kathleen B. Bamford. © 2012 John Wiley & Sons, Ltd.

Fomites

Any inanimate object may be contaminated with organisms and act as a vehicle (fomite) for transmission. This is important for doctors performing procedures on patients with instruments that might be contaminated and transmit infection (e.g. MRSA on stethoscopes).

Water supply

The water supply in a hospital is a complex system, supplying water to wash-hand basins and showers, central heating and air-conditioning systems. *Legionella* spp. may colonize the system in redundant areas of pipework and cooling-tower systems are a particular risk. When systems fail, organisms such as *Legionella* can be transmitted by the air-conditioning system. To reduce this risk, hot-water supplies should be maintained at a temperature above 45 °C and cold-water supplies below 20 °C.

The host

Hospital patients are susceptible to infection as a result of underlying illness or treatment, for example patients with leukaemia or those receiving cytotoxic chemotherapy. Age and immobility may predispose to infection; ischaemia may make tissues more susceptible to bacterial invasion.

Medical activities

Intravenous access

This is the most frequent source of healthcare-associated bacteraemia. The risk of infection from any intravenous device increases with the length of time it remains in position. Having broken the skin's integrity, it provides a route for invasion by skin organisms such as *Staphylococcus aureus*, *Staphylococcus epidermidis* and *Corynebacterium jeikeium*. Signs of inflammation at the puncture site may be the first evidence of infection. Cannula-related infection can be complicated by septicaemia, endocarditis and metastatic infections (e.g. osteomyelitis). The risk of sepsis can be reduced by aseptic technique at insertion, as well as the choice of device, for instance those without side ports and dead spaces. Maintaining an adequate dressing and ensuring good staff hygiene when they are working with the device are equally important. The cannula site should be regularly inspected and this is particularly important in unconscious patients. Peripheral lines should be re-sited every 48 h; central and tunnelled lines should be changed when there is evidence of infection.

Urinary catheters

Indwelling urinary catheters bypass the normal defences and provide a route for ascending infection into the bladder. Risks can be minimized by aseptic technique when the catheter is inserted and handled.

Respiratory

Intubation bypasses the defences of the respiratory tract. Postoperative pain, immobility and the effects of anaesthesia predispose to pneumonia by reducing coughing. Respiratory infections with resistant Gram-negative organisms that originate from the hospital environment may occur. Inhalation of oral contents is reduced by raising the head of the bed in seriously ill patients.

Surgery

Surgical patients often have other health problems that are unrelated to their surgical complaint (e.g. asthma or diabetes mellitus), which may predispose them to infection. Surgery is traumatic and carries a risk of infection (e.g. especially wound infections).

Complications of the procedure may increase risks (e.g. postoperative ischaemia). Operation length and complexity influence the risk of infection, as does the skill of the surgeon: the less tissue damage occurs at an operation, the lower the risk of infection. The preoperative period should be short to reduce the risk of acquiring resistant hospital organisms and elective surgery should be postponed in patients with active infection (e.g. chest infections).

To minimize the risk of infection during an operation, theatres are supplied with a filtered air supply. Staff movement during procedures should be limited to reduce air disturbance. Changing clothing reduces transmission of organisms from the wards. Impervious materials reduce contamination from the skin of the surgical team but are uncomfortable to wear. Some hospitals provide ventilated air-conditioned suits for surgical teams performing prosthetic joint surgery.

Antibiotic prophylaxis

Antibiotic prophylaxis reduces postoperative infection rates. Antibiotics should be bactericidal, have sufficient penetration for the required site and be chosen on the basis of the operation type – see below. There is no evidence that continuing prophylaxis beyond 48 h is beneficial.

'Clean' involving the skin or a normally sterile site (e.g. a joint). Antibiotics are unnecessary unless a prosthetic device is inserted.

'Contaminated' where a viscus with a normal flora is breached. Antibiotics such as metronidazole and a second-generation cephalosporin for large-bowel surgery; a cephalosporin alone is satisfactory in surgery of the upper gastrointestinal or biliary tracts where anaerobes are rarely implicated.

'Infected' operations are those in which surgery is required to deal with an already infected situation, such as drainage of an abscess or repair of a perforated diverticulum. Parenteral antibiotics to treat the likely infecting organisms should be prescribed.

11 Immunization

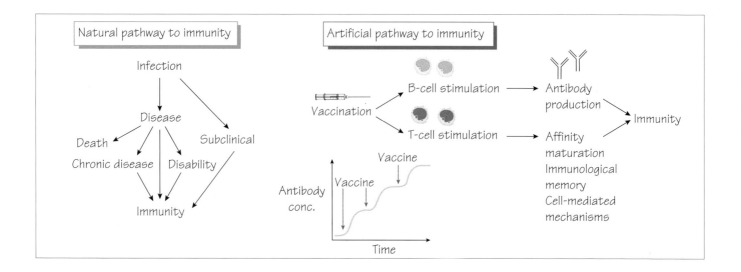

Immunization

Vaccination has virtually eradicated many infectious diseases (e.g. polio and diphtheria) and has made smallpox extinct. Immunization may be achieved passively, by administration of an immunoglobulin preparation, or actively, by vaccination.

Passive immunization

Immunoglobulin, prepared from pooled plasma, may provide short-term protection against infections, such as measles, if given after exposure. This is used for immunocompromised individuals exposed to specific infections. It is also used to protect patients with congenital immunodeficiency. Specific products are prepared from hyperimmune donors and used for postexposure management (e.g. hepatitis B, varicella zoster and tetanus).

Immunization

Immunization aims to generate **immunity** without the complications associated with the natural infection and is achieved by administering a **vaccine** that generates immunity without harm. There are different types of vaccines.
• Whole organisms (bacteria or viruses, e.g. influenza).
• Isolated antigenic components (acellular, e.g. meningitis C vaccine).
• Live attenuated vaccines that contain strains with reduced/absent pathogenicity (e.g. MMR). Live vaccines can cause disease in immunocompromised patients and are avoided in pregnancy because of the risk of fetal infection.
• Toxins of tetanus and diphtheria are inactivated (toxoids) so that they are non-toxic but fully immunogenic.

Vaccine immunogenicity can be increased by adding adjuvants or by conjugation with proteins (e.g. against *Haemophilus influenzae* type b). Genetic engineering can be used to create acellular vaccines safely (e.g. against hepatitis B).

Immunization programmes can be used to eradicate a pathogen (e.g. smallpox, which is complete, and polio, which is nearing completion), but this is only achievable where there is no non-human reservoir of the infection. Elimination can be achieved where for instance the disease has largely disappeared, but the organism remains in humans, animal hosts or the environment (e.g. clinical diphtheria and tetanus). Containment is achieved where the burden of disease or transmission is reduced significantly (e.g. pertussis).

Universal immunization is adopted for most childhood infections (see diagram of vaccination schedules). Selective vaccination is used for those at risk from a disease that is otherwise rare (e.g. hepatitis B in healthcare workers). Immunization programmes are updated regularly to take account of changing disease epidemiology and new vaccine availability.

Vaccine safety

All clinical interventions carry some degree of risk. In determining vaccine policy the risks from the natural disease must significantly outweigh the risks of vaccination. The safety of any vaccine is of paramount importance. Extensive trials are carried out to evaluate safety and strict 'good manufacturing practice' legislation governs the production of vaccines. When a vaccine is given, records are kept of the circumstances of administration and the vaccine lot number. A number of highly publicized false vaccine scares have undermined health professional and public confidence in a number of childhood vaccines. This in turn has reduced uptake and resulted in deaths and lifelong morbidity as a consequence of re-emergence of diseases such as whooping cough and measles.

Up-to-date information about immunization can be obtained in the UK from the Health Protection Agency (www.hpa.org.uk/web/HPAweb&Page./1204031508623); in the USA from the Infectious Diseases Society of America and the Centers for Disease Control and Prevention (http://www.idsociety.org/ and http://www.cdc.gov/); in the Pacific from http://immunise.health.gov.au/publications.htm; and from the World Health Organization (http://whqlibdoc.who.int/publications/2005/9241580364_chap6.pdf).

Travellers should take specialized advice about the vaccinations needed for their journeys (e.g. yellow fever vaccine).

Medical Microbiology and Infection at a Glance, Fourth Edition. Stephen H. Gillespie, Kathleen B. Bamford. © 2012 John Wiley & Sons, Ltd.

Age ► Vaccine ▼	Birth	1 month	2 months	3 months	4 months	6 months	12–13 months	18 months	24 months	3.5–5 years	12–13 years	13–18 years	19–49 years	>50 years
Diptheria												1 dose every 10 years →		
Tetanus														
Pertussis														
Polio														
Haemophilus influenza B														
Meningitis C														
Pneumococcal											1 dose if vaccination history unclear, 2 doses if occupational/other indications			
Measles, mumps, rubella														
Human papiloma virus											Girls only			
Influenza						1 annual dose								
Hepatitis A										2 doses 0, 6–12 months for those with medical or other indications				
Hepatitis B	3 doses 0, 4 weeks, 6 months for those with medical, behavioural, occupational and other indications													
BCG		In high risk districts												
Varicella						2 doses 0, 4–8 weeks if susceptible for those with medical, occupational or other indications								

Legend: ☐ Child hood immunisation schedule ☐ Risk groups ☐ All

Vaccine ▼ Age ►	19–49 years	50–64 years	65 years and older
Tetanus, diphtheria (Td)	1 dose booster every 10 years		
Influenza	1 dose annually for persons with medical and occupational indications, or household contacts of persons with indications	1 annual dose	
Pneumococcal (polysaccharide)	1 dose for persons with medical or other indications (1 dose revaccination for immunosuppressive conditions)		1 dose for unvaccinated persons / 1-dose revaccination
Hepatitis B	3 doses (0, 4 weeks, 6 months) for persons with medical, behavioral, occupational, and other indications		
Hepatitis A	2 doses (0, 6–12 months) for persons with medical and other indications		
MMR	1 dose if measles, mumps, or rubella vaccination history is unreliable; 2 doses for persons with occupational, geographic, and other indications		
Varicella	2 doses (0, 4–8 weeks) for persons who are susceptible		
Meningococcal (polysaccharide)	1 dose for persons with medical, geographic, or occupational indications		

Legend: ☐ For all persons in this age group ☐ Catch-up on childhood vaccinations ☐ For persons with medical exposure indications

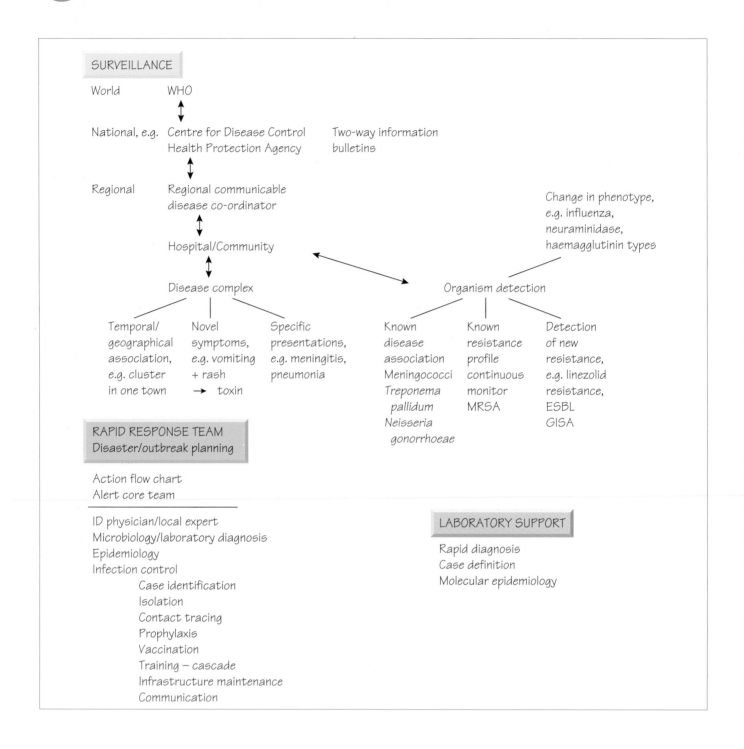

The incidence of an infection varies with changes in pathogen virulence and in the immunity of the host population. Many infectious diseases have a characteristic pattern. For example, meningococcal infection in the 'meningitis belt' of Africa is endemic (cases occur all of the time), but the incidence of the disease will sometimes rise to 1000 per 100 000 of the population – an epidemic.

The words 'outbreak' and 'epidemic' are often used interchangeably, although an outbreak can be thought of as a localized epidemic. Pandemics occur when an epidemic spreads worldwide. Examples of pandemic diseases include influenza after the organism undergoes an antigenic shift (see Chapter 34), plague (Chapter 22) and cholera (Chapter 25).

The speed at which an infection spreads through a community depends on the mechanism of transmission (see Chapter 8). For example, organisms that spread by the respiratory route can spread more rapidly than organisms that spread by the sexual route. The **infectivity** of a pathogen is defined by the speed at which an organism spreads in a community (measles, for example, is highly infectious, whereas mumps is less so); it can be quantified using the 'intrinsic reproductive number', which is the average number of secondary cases arising from a single case in a totally susceptible population. Initially the number of cases continues to rise, but the number of susceptible individuals falls because of death or the development of immunity. Then as the proportion of susceptible individuals falls, the number of new cases (incidence) starts to decline. Mathematical models can be used to predict the outcome of an epidemic and to indicate control methods.

Emerging infections

New or emerging infections are important because 'new' diseases may not be recognized immediately, so may remain undiagnosed. Emerging diseases fall into three broad categories:
- new pathogens;
- established pathogens invading new territories;
- re-emergence of a previously controlled organism.

New pathogens

Several of the most important emerging infections have arisen because they are genuinely new infections. The most important example of this is the human immunodeficiency virus (HIV), which is thought to have emerged from simian immunodeficiency virus (SIV) by jumping species from chimpanzees to humans in central Africa over 50 years ago (see Chapter 46). Severe acute respiratory syndrome (SARS) was caused by a novel coronavirus that jumped into humans in southern China; fortunately this infection was controlled rapidly. Invasion of forest habitats has brought humans into contact with *Trypanosoma cruzi*, which lead to infections.

New territory

Climatic change or changes in population centres may permit an organism to invade new territory. For example, infection with West Nile virus is spreading in the USA at present, entering some states for the first time. Climate change may allow vectors to change their range and permit diseases like malaria to invade/re-invade new territories.

Re-emergence of previously uncommon infections

Effective treatment and organized control programmes had been reducing tuberculosis; however, migration and synergy with HIV has now increased the number of cases globally. Loss of confidence in MMR resulted in a reduction in vaccine uptake, which resulted in localized outbreaks of measles that showed that many infectious diseases await their opportunity to re-emerge. The development of multiple-drug resistance may allow other organisms to re-emerge (e.g. multidrug-resistant tuberculosis [MDRTB]).

Changes in agriculture and food production

The spread of organisms in livestock herds or flocks, such as strains of *Salmonella enteritidis*, may provide new opportunities for contamination of products in the food chain. The widespread use of pre-prepared chilled food has increased the risk of listeriosis.

Changes in the virulence or transmissibility of a pathogen

Some pathogens may emerge in particular niches because they are more virulent or more easily transmitted. *Clostridium difficile* 027 has caused hospital outbreaks in Canada, the USA and the UK. Multiple drug-resistant *Acinetobacter baumanni* has caused outbreaks in intensive care units. Epidemic MRSA EMRSA-15 and ERMSA-16 have caused outbreaks in many settings and are now the predominant strains of MRSA worldwide.

Bioterrorism

Political uncertainty and terrorist activity raise the risk that microbes could be used as biological weapons. One bioterrorism incident using anthrax occurred in the USA and resulted in four deaths and the need to create extensive control measures. Other agents of concern include smallpox, tularaemia, plague and a number of viral haemorrhagic fever viruses. Healthcare professionals need to know about the signs of these unusual infections and how to communicate concerns to the relevant agencies so that changing patterns can be identified quickly and interventions rapidly deployed (see Chapter 9).

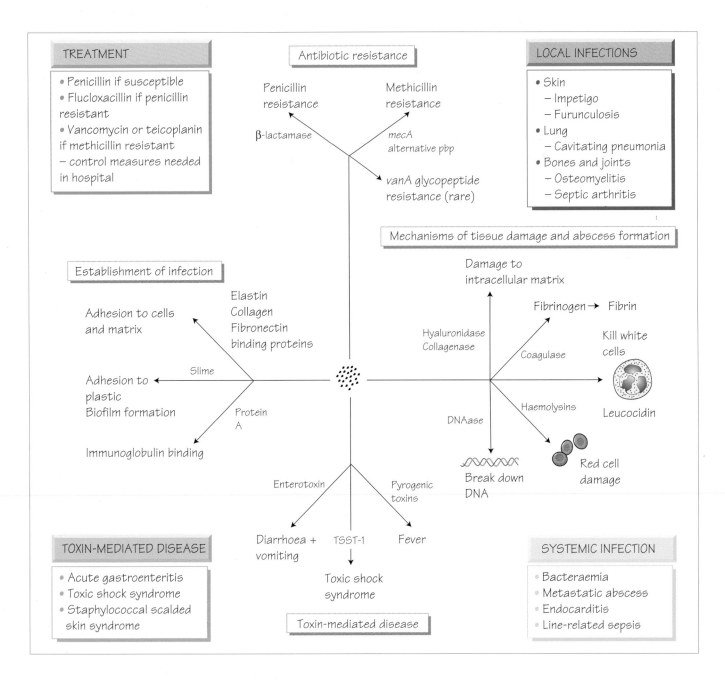

TREATMENT

- Penicillin if susceptible
- Flucloxacillin if penicillin resistant
- Vancomycin or teicoplanin if methicillin resistant
  – control measures needed in hospital

Antibiotic resistance

Penicillin resistance

Methicillin resistance

β-lactamase

mecA alternative pbp

vanA glycopeptide resistance (rare)

LOCAL INFECTIONS

- Skin
  – Impetigo
  – Furunculosis
- Lung
  – Cavitating pneumonia
- Bones and joints
  – Osteomyelitis
  – Septic arthritis

Mechanisms of tissue damage and abscess formation

Establishment of infection

Adhesion to cells and matrix

Elastin
Collagen
Fibronectin
binding proteins

Slime

Adhesion to plastic
Biofilm formation

Protein A

Immunoglobulin binding

Damage to intracellular matrix

Hyaluronidase
Collagenase

Fibrinogen → Fibrin

Coagulase

Kill white cells

DNAase

Haemolysins

Leucocidin

Break down DNA

Red cell damage

Enterotoxin

Pyrogenic toxins

Diarrhoea + vomiting

TSST-1

Fever

Toxic shock syndrome

Toxin-mediated disease

TOXIN-MEDIATED DISEASE

- Acute gastroenteritis
- Toxic shock syndrome
- Staphylococcal scalded skin syndrome

SYSTEMIC INFECTION

- Bacteraemia
- Metastatic abscess
- Endocarditis
- Line-related sepsis

Staphylococci are part of the normal flora and important human pathogens. There are more than 26 species but only a few are pathogenic. *Staphylococcus aureus* is the most invasive species, which can be differentiated from other species by its possession of the enzyme coagulase.

Staphylococcus aureus

Asymptomatic carriage of *S. aureus* is found in up to 40% of healthy people, in the nose, skin, axilla or perineum. This is impor-tant in healthcare workers especially if they carry an invasive or resistant strain (e.g. MRSA).

Pathogenesis

Staphylococcus aureus has many potential pathogenicity determi-nants such as coagulase, which catalyses the conversion of fibrino-gen to fibrin thereby providing protection.

Medical Microbiology and Infection at a Glance, Fourth Edition. Stephen H. Gillespie, Kathleen B. Bamford. © 2012 John Wiley & Sons, Ltd. Published 2012 by John Wiley & Sons, Ltd.

Determinant	Activity	Effect
Coagulase	Converts fibrinogen to fibrin	May form protective barrier
Adhesion molecules	Bind fibronectin and collagen	Assist adherence
Lytic enzymes	Lipase	Breaks down host tissue
Protein toxins	Panton–Valentine leucocidin (PVL) toxin	Tissue damage
	Toxic shock syndrome toxin (TSST)	Shock and toxicity
	Enterotoxins	Diarrhoea
Biofilm formation	Slower growth in an extracellular matrix	Difficult to treat with antibiotics, adhere to plastics

Clinical importance

Staphylococcus aureus causes a wide range of infectious syndromes.
• Primary skin infection – impetigo, which is transmitted from person to person.
• Secondary skin infections – associated with eczema, surgical wounds, intravenous devices, burns.
• Pneumonia – rare, but may follow influenza and progress rapidly with cavity formation (see Chapter 34).
• Endocarditis – can be rapid and destructive; associated with intravenous drug misuse or colonization of intravenous devices (see Chapter 48).
• Osteomyelitis (see Chapter 52).
• Septic arthritis (see Chapter 52).

Laboratory diagnosis

Staphylococcus aureus grows readily on most laboratory media. Selective medium contains high salt, to which *S. aureus* is relatively tolerant. Phenotypic identification depends on demonstrating coagulase, catalase enzymes and typical 'cluster of grapes' morphology on Gram stain. Typing by molecular means can support interventions to control outbreaks.

Antibiotic susceptibility

The history of the susceptibility of *S. aureus* is a lesson in the history of antimicrobial chemotherapy.
1 It was initially susceptible to penicillin, but strains that produced β-lactamase soon predominated, so methicillin and related agents (e.g. flucloxacillin) were introduced and replaced penicillin.
2 Methicillin-resistant *S. aureus* (MRSA) emerged. Resistance is caused by possession of the *mecA* gene, which codes for a penicillin-binding protein that binds the drug less well. Glycopeptides, such as vancomycin or teicoplanin, started to be required for these strains.
3 Intermediate or heteroresistance to glycopeptides emerged as an increasing issue and fully glycopeptide-resistant strains (GRSA)

have now been described, resistance being mediated by the *vanA vanB* genes acquired from enterococci.

Other antibiotics that remain effective include linezolid, aminoglycosides, erythromycin, clindamycin, fusidic acid, chloramphenicol and tetracycline.

In methicillin-sensitive strains, first- and second-generation cephalosporins are effective. Fusidic acid may be given with another agent; it is often given in bone and joint infections (see Chapter 52) because of its tissue penetration.

Prevention and control

Staphylococcus aureus spreads by airborne transmission and via the hands of healthcare workers. Patients colonized or infected with MRSA or GRSA should be isolated in a side room with wound and enteric precautions (see Chapter 10). Staff may become carriers and disseminate the organism widely in the hospital environment. Carriage may be eradicated by using topical mupirocin and chlorhexidine.

Staphylococcus epidermidis

Staphylococcus epidermidis is the most important of the coagulase-negative staphylococci (CoNS). Once dismissed as contaminants, these organisms are now recognized as pathogens if conditions favour their multiplication.

Clinical importance

Staphylococcus epidermidis causes infection of intravenous cannulae, long-standing intravascular prosthetic devices, ventriculoperitoneal shunts and prosthetic joints, which may lead to bacteraemia or endocarditis and require the removal of the prosthesis. Biofilm production contributes to their pathogenicity.

Laboratory diagnosis

Staphylococcus epidermidis grows readily on laboratory media; coagulase is not produced. Speciation is by biochemical testing DNA restriction patterns or other molecular techniques may be needed to determine whether strains are identical. *S. epidermidis* and other CoNS are common contaminants in blood cultures, requiring careful evaluation of their clinical significance.

Antibiotic susceptibility

This group of organisms is uniformly susceptible to vancomycin and usually to teicoplanin. They may also be susceptible to any of the other agents used for *S. aureus* infection, but this is unpredictable. Treatment must be guided by *in vitro* testing.

Staphylococcus haemolyticus

Less common than *S. epidermidis*, *Staphylococcus haemolyticus* causes a similar disease pattern. It differs from *S. epidermidis* in that it causes haemolysis on blood agar. More importantly, it is naturally resistant to teicoplanin; significant infections require vancomycin therapy.

Staphylococcus saprophyticus

These CoNS are a common cause of urinary tract infection in young women. They can be rapidly distinguished from other species by their resistance to novobiocin.

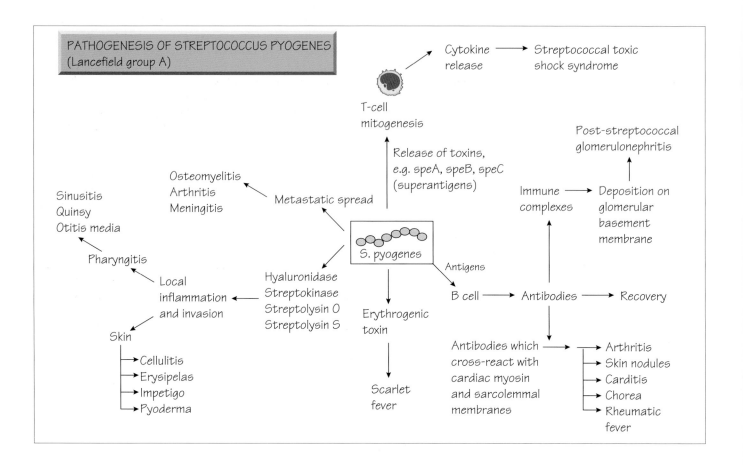

PATHOGENESIS OF STREPTOCOCCUS PYOGENES (Lancefield group A)

The main characteristics of streptococci include the following.
- They are Gram-positive cocci arranged in pairs and chains.
- They are fastidious facultative anaerobes.
- They require rich blood-containing media.
- They may be cultured from the site of infection (throat, wound, etc.) or in blood cultures.
- Colonies are distinguished by haemolysis: complete (β) or incomplete (α) and none (γ).

Streptococcus pyogenes

Streptococcus pyogenes is carried asymptomatically in the pharynx of 5–30% of the population, more commonly in children. It is transmitted by the aerosol route and by contact.

Pathogenesis

Streptococcus pyogenes is surrounded by the M protein antigen, which reduces leucocyte phagocytosis. Antibodies to M proteins are protective but only against infection with the same M type and there are multiple M types. They adhere via fibronectin receptors. Other features are:
- toxins are important in their pathogenicity:
  ◦ erythrogenic toxin associated with scarlet fever;
  ◦ pyrogenic exotoxins A, B and C associated with toxic shock;

- production of *S. pyogenes* cell envelope proteinase (SpyCEP), which degrades IL-8 and other cytokines, thereby retarding neutrophil activation;
- ability to invade and survive intracellularly making them difficult to eradicate with penicillin;
- production of degrading enzymes (immunoglobulin proteases, hyaluronidase and collagenases).

Clinical presentation

Streptococcus pyogenes, which is an important cause of mortality worldwide, has the following associations:

1 Invasive disease characterized by rapid onset, local tissue destruction and rapid spread within the tissues. Systemic toxicity is common and associated with toxin production. Syndromes include:
- pharyngitis – the most common bacterial cause (see Chapter 50);
- skin infection – erysipelas, impetigo, cellulitis, wound infections and rarely, necrotizing fasciitis or pneumonia (see Chapter 57);
- puerperal sepsis (see Chapter 45);
- complications such as septicaemia and metastatic infections (e.g. osteomyelitis);
- severe toxicity:

Medical Microbiology and Infection at a Glance, Fourth Edition. Stephen H. Gillespie, Kathleen B. Bamford. © 2012 John Wiley & Sons, Ltd.

- erythrogenic toxin causes scarlet fever;
- pyrogenic toxin-producing strains are associated with streptococcal shock and have a high mortality due to multiple organ failure.

2 Postinfectious immune-mediated diseases which include rheumatic fever, glomerulonephritis and erythema nodosum, are thought to be immune-mediated because antibodies to bacterial structures cross-react *with* host tissues. Rheumatic fever, now uncommon in developed countries, is a major cause of long-term morbidity and mortality, particularly in areas of poverty and malnutrition.

Prevention and control

Streptococcus pyogenes can spread rapidly in surgical and obstetric wards; infected or colonized patients should be isolated in a side room until 48h after initiation of effective antibiotics. Prompt treatment prevents secondary immune disease (e.g. rheumatic fever). Benzylpenicillin is the treatment of choice and resistance has never been reported. Amoxicillin may be used for oral therapy in less severe infections. Macrolides are an alternative for patients with allergy.

Streptococcus agalactiae

Streptococcus agalactiae (group B streptococcus) is a commensal in the gut and female genital tract. It causes:
- early perinatal pneumonia or septicaemia;
- later perinatal meningitis;
- puerperal sepsis.

The polysaccharide antiphagocytic capsule is the main pathogenicity determinant. Prophylactic therapy to prevent neonatal disease can be given to those mothers in labour who are febrile, known to be colonized or who have previously had an affected child. There is currently no vaccine available.

Clinical features and diagnosis

Infected neonates may initially lack the classical clinical signs of sepsis, such as fever and the bulging fontanelle of meningitis. A chest X-ray may demonstrate pneumonia, and specimens of blood, CSF, amniotic fluid and gastric aspirate should be cultured. Antigen detection tests are available and can be applied to body fluids for rapid diagnosis.

Treatment and prevention

Neonatal group B streptococcal sepsis requires empirical therapy that includes a penicillin and an aminoglycoside. Perinatal penicillin can prevent invasive infection but should be targeted at babies who are at high risk.

Enterococcus spp.

Enterococci possess a group D carbohydrate cell wall antigen and can exhibit all three types of haemolysis (see above). They are principally commensals of the bowel but may cause disease if they become established at other sites. Of more than 12 species, *Enterococcus faecalis* and *E. faecium* are the most common members to act as human pathogens, causing urinary tract infection, wound infection and endocarditis. Enterococci are emerging as hospital pathogens, with some species (e.g. *E. faecium*) now resistant to commonly used antibiotics. Most strains are sensitive to ampicillin/amoxicillin; however, resistance levels are increasing. Some enterococci have developed resistance to glycopeptides by the acquisition of an alternative peptidoglycan transpeptidation enzyme (the van A or van C system) that alters the usual Ala–D-Ala cross-linkage so it is not glycopeptide susceptible. Such strains may require therapy with drugs such as linezolid, daptomycin or pristinamycin.

15 Streptococcus pneumoniae, other Gram-positive cocci and the alpha-haemolytic streptococci

S. pneumoniae infections

Sinusitis

Otitis media
Mastoiditis

Pneumonia
Empyema

Endocarditis
(rare)

Meningitis

Bacteraemia
Systemic disease
Shock
GI disturbance
CRP

Other streptococci

Streptococcus agalactiae (group B)	Neonatal pneumonia, septicaemia, meningitis, rarely adult infection
Streptococcus equi, S. equisimilis, S. dysgalactiae, S. zooepidemicus (group C)	Endocarditis, skin infection, pharyngitis
Enterococci (group D) E. faecalis, faecium, E. bovis	UTI, endocarditis, intra-abdominal polymicrobial infection, infection in immunocompromised, ≠ resistance to glycopeptides
S. milleri group	
S. anginosus-intermedius-constellatus	Metastatic abscesses, periodontal sepsis
S. oralis, S. sanguis, S. mitis	Endocarditis
S. mutans	Dental caries

Streptococcus pneumoniae

Streptococcus pneumoniae (or pneumococcus) is a Gram-positive coccus seen in pairs, which is typically α-haemolytic, but can be variable.

Pathogenesis

Streptococcus pneumoniae has a polysaccharide capsule that protects it from phagocytosis. There are over 90 highly antigenic capsular serotypes and antibodies to specific types are protective.

Pathogenicity features include:
- pro-inflammatory cell wall components (e.g. C-polysaccharide, F-antigen);
- IgA2 protease;
- pneumolysin, a cytotoxin that stimulates immune responses;
- adhesins that bind to cell surface carbohydrates (e.g. choline binding protein A, pneumococcal surface protein A [PspA]);
- tissue damaging enzymes (e.g. neuraminidase, hyaluronidase).

Epidemiology

Humans are the only host of *S. pneumoniae*. Carriage, which is usually asymptomatic, is most common in the young or smokers and is associated with overcrowding. Serotypes vary with country, time and subject group.

Children under 1 year of age are vulnerable to acute pneumonia. Complement deficiency, agammaglobulinaemia, HIV infection, smoking, alcoholism and splenectomy predispose to severe infection. The bacteria are able to adhere to pneumocytes and invade the bloodstream by hijacking the platelet-aggregating factor receptor pathway and produce complement-mediated damage to the alveolus through the action of pneumolysin.

Clinical features

- Acute otitis media, sinusitis and acute pneumonia are the most common infections.
- Pneumococci cause between 50 and 75% of cases of community-acquired pneumonia, up to 25–30% of which may develop bacteraemia.

Medical Microbiology and Infection at a Glance, Fourth Edition. Stephen H. Gillespie, Kathleen B. Bamford. © 2012 John Wiley & Sons, Ltd.

- Bacteraemia is an important complication with a high mortality, despite treatment (see Chapter 47).
- Direct or haematogenous spread can give rise to meningitis, which has a high mortality and is associated with brain damage. This is now the commonest cause of meningitis in adults over 40 and the second commonest cause in children from populations that have been vaccinated against *Haemophilus influenzae* type b (Hib).
- Pneumococcus rarely causes cellulitis, abscesses, peritonitis and endocarditis.
- The mortality and incidence of sequelae are high.

Antibiotic susceptibility and treatment
Once universally susceptible to penicillin, significant numbers of *S. pneumoniae* have developed resistance through a genetically modified penicillin-binding protein gene (see Chapter 7), and penicillin-resistant clones have spread internationally. *S. pneumoniae* is also susceptible to erythromycin, cephalosporins, tetracycline, rifampicin and chloramphenicol, but multiple drug resistance is growing. Penicillin is the treatment of choice for respiratory infection but third generation cephalosporins are used for meningitis if it is caused by less sensitive strains. Where high-level penicillin resistance occurs, a glycopeptide (usually vancomycin) should be added.

Prevention and control
A conjugate vaccine incorporating up to 13 capsular serotypes has been introduced and is highly immunogenic in young children. Whilst this has led to a decline in invasive pneumococcal disease in children and adults, there is some evidence that serotypes not included in the vaccine are increasing.

Alpha-haemolytic streptococci
There are a wide range of streptococci found in the oropharynx, which can occasionally cause disease. Some of these species are closely related to *S. pneumoniae*.

Infective endocarditis
The α-haemolytic streptococci (*S. oralis, S. sanguis, S. mutans* and *S. salivarius*) cause 40–60% of community-acquired native-valve endocarditis. Infection may be of dental origin; while good evidence is lacking, prophylaxis is recommended for at-risk patients undergoing bacteraemia-inducing dental procedures such as extraction or deep scaling (see Chapter 48). *Streptococcus bovis* bacteraemia and endocarditis is associated with underlying bowel malignancy. Occasionally endocarditis is caused by nutritionally deficient (pyridoxine-dependent) streptococci that can be missed on culture.

Metastatic abscesses
The '*Streptococcus milleri*' group of organisms (*S. anginosus, S. intermedius* and *S. constellatus*) colonize the mouth and gut. They are sometime responsible for metastatic infection, causing brain, lung or liver abscesses often as part of a mixed infection with obligate anaerobes. Isolation of a member of the '*S. milleri*' group should prompt a thorough search for an occult abscess.

Other gram-positive cocci
A number of other Gram-positive cocci such as *Leuconostoc* and *Pediococcus* are occasionally associated with infections, particularly in immunocompromised individuals.

Alloiococcus otitidis
Alloiococcus otitidis is a slow-growing Gram-positive coccus that produces lactic acid and has been associated with chronic otitis media with effusion in children, particularly in the chronic phase although its pathogenicity is not certain.

Streptococcus pseudopneumoniae
This organism is genetically closely related to *S. pneumoniae* but does not have a capsule and may lack some of the common pathogenicity determinants. It has been associated with isolation from patients with chronic obstructive pulmonary disease.

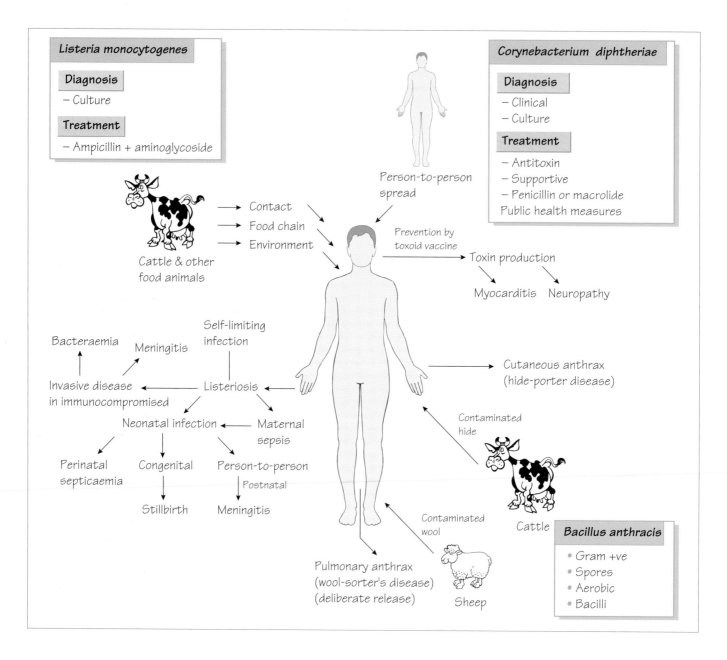

Listeria monocytogenes

Diagnosis
– Culture

Treatment
– Ampicillin + aminoglycoside

Cattle & other food animals

→ Contact
→ Food chain
→ Environment

Person-to-person spread

Corynebacterium diphtheriae

Diagnosis
– Clinical
– Culture

Treatment
– Antitoxin
– Supportive
– Penicillin or macrolide
Public health measures

Prevention by toxoid vaccine

Toxin production

Myocarditis   Neuropathy

Bacteraemia   Meningitis

Self-limiting infection

Invasive disease in immunocompromised ← Listeriosis

Cutaneous anthrax (hide-porter disease)

Contaminated hide

Neonatal infection ← Maternal sepsis

Perinatal septicaemia   Congenital   Person-to-person
                                         Postnatal
Stillbirth   Meningitis

Cattle

Contaminated wool

Pulmonary anthrax (wool-sorter's disease) (deliberate release)

Sheep

Bacillus anthracis
• Gram +ve
• Spores
• Aerobic
• Bacilli

Listeria

Listeria are important Gram-positive organisms that can grow at low temperatures (4–10 °C); *Listeria monocytogenes* is associated with human disease.

Epidemiology

Listeria spp. are found in soil or animal faeces and can contaminate foodstuffs. Infection follows consumption of contaminated food; inadequately pasteurized foods and contaminated foods stored in the fridge are at risk.

Clinical features

Listeria monocytogenes infection is usually a mild, self-limiting, infectious mononucleosis-like syndrome. Rarely acute pyogenic meningitis, bacteraemia or encephalitis can develop and these conditions carry a high mortality rate, particularly in patients with

Medical Microbiology and Infection at a Glance, Fourth Edition. Stephen H. Gillespie, Kathleen B. Bamford. © 2012 John Wiley & Sons, Ltd.

reduced cell-mediated immunity. Bacteraemia occurring in pregnancy is associated with intrauterine death, premature labour and neonatal infection similar to that seen with group B streptococci (see Chapter 45).

Laboratory diagnosis

Diagnosis is by culture on simple laboratory media and further identification is made by biochemical testing. Typing is achieved by multilocus sequence typing (MLST).

Management

Listeria spp. are susceptible to ampicillin and gentamicin but resistant to the cephalosporins, penicillin and chloramphenicol. Patients with symptoms of meningitis, in whom listeriosis is a possible diagnosis, should receive a drug regimen that incorporates ampicillin.

Prevention and control

Listeriosis is controlled by food hygiene, effective refrigeration and adequate reheating of pre-prepared food. Individuals at particular risk, such as pregnant women and immunocompromised patients, should avoid high-risk foods.

Corynebacterium spp.

Corynebacterium jeikeium

This organism, which is naturally resistant to most antibiotics except vancomycin, colonises prostheses and intravenous lines causing infection and bacteraemia, usually in immunocompromised individuals.

Other corynebacteria and related organisms

Rarely, *Corynebacterium ulcerans* can carry the phage that encodes diphtheria toxin and may cause a diphtheria-like pharyngitis. *Corynebacterium pseudotuberculosis* may cause suppurative granulomatous lymphadenitis. *Rhodococcus equi* has been associated with a severe cavitating pneumonia in patients with acquired immune deficiency syndrome (AIDS).

Non-tuberculous mycobacteria

Different species may cause localized or disseminated disease in immunocompromised patients. Some may infect prosthetic devices.

Mycobacterium avium–intracellulare complex

The mycobacterium avium–intracellulare complex (MAIC) includes *Mycobacterium avium*, *M. intracellulare* and *M. scrofulaceum*, some being natural pathogens of birds, others being environmental saprophytes. They are a common cause of mycobacterial lymphadenitis in children, also causing osteomyelitis in immunocompromised patients and chronic pulmonary infection in the elderly. In the advanced stages of AIDS, they cause disseminated infection and bacteraemia. MAIC is naturally resistant to many antituberculosis agents and treatment with multidrug regimens that include rifabutin, clarithromycin and ethambutol is usually required. Lymphadenitis may require surgery.

Mycobacterium kansasi, Mycobacterium malmoense and *Mycobacterium xenopi*

These species cause an indolent pulmonary infection that resembles tuberculosis in individuals predisposed by chronic lung disease that has caused deranged pulmonary anatomy (e.g. bronchiectasis, silicosis and obstructive airways disease). Initial therapy with standard drugs may have to be adjusted following the results of bacterial identification and susceptibility tests.

Bacillus

The spores produced by these Gram-positive aerobic bacilli allow them to survive in adverse environmental conditions.

Bacillus anthracis

Bacillus anthracis is a soil organism that, under certain climatic conditions, multiplies to cause anthrax in herbivores. Humans can become infected from contaminated animal products. Pathogenicity depends on three bacterial antigens: the 'protective antigen', the oedema factor (both of which are toxins) and the antiphagocytic poly D-glutamic acid capsule. Inoculation of *B. anthracis* into minor skin abrasions produces a necrotic, oedematous ulcer with regional lymphadenopathy. Inhalation of anthrax spores develops into fulminant pneumonia and septicaemia. An outbreak in 2001 that was caused by deliberate release of the organism has led to the recognition of anthrax as an agent of bioterrorism. The spores of the organisms are prepared in a way that makes them readily aerosolized so that they spread rapidly, infecting many by the respiratory route.

Rapid diagnosis by nucleic acid amplification test (NAAT) is available for potential bioterrorism exposure. The definitive diagnosis must be made in a laboratory equipped for and specialized in handling this organism. Treatment is with a penicillin, fluoroquinolone, erythromycin or tetracycline. Anthrax is prevented by animal vaccination, sporicidal treatment of animal products and vaccination of humans at high risk. Antibiotic prophylaxis is used to prevent disease associated with known exposure. Vaccines are available for military and related personal when biological agents are a risk.

Bacillus cereus

Bacillus cereus produces a heat-stable toxin. Typically, it multiplies in parboiled rice and other contaminated food products, causing a self-limiting food poisoning: vomiting occurs 6h after exposure, followed by diarrhoea (18h).

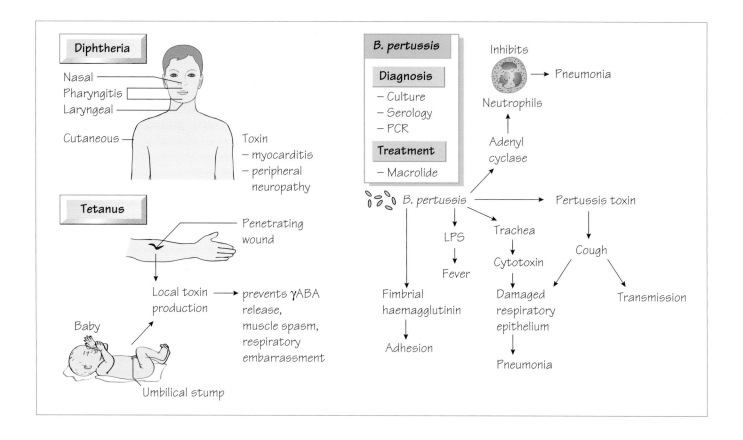

These three organisms are from widely differing taxonomic groups but are linked by being important diseases of childhood that are mediated by toxins and can be prevented by childhood immunization (see Chapter 11).

Corynebacterium diphtheriae

Pathogenesis

Diphtheria is caused by *Corynebacterium diphtheriae* that contain a bacteriophage encoding diphtheria toxin. The toxin kills cells by interrupting protein synthesis, acting on the myocardium to cause myocarditis and on the peripheral nervous system to cause neuropathy and paralysis. The severity of infection is directly related to the degree of toxin production. Cutaneous infection is often asymptomatic.

Corynebacterium diphtheriae is transmitted by the respiratory route or following direct contact with cutaneous lesions.

Clinical features and management

Infection of the skin, nasopharynx or larynx may occur; the severity of disease is related to the extent of the infection. A rare cause of sore throat, it causes inflammation and necrosis giving a green–black 'pseudomembrane' on the posterior wall of the pharynx, which can cause respiratory obstruction. Management is based on isolation and treatment with antitoxin and erythromycin. Intensive care support may be required.

Laboratory diagnosis

Corynebacterium diphtheriae is isolated using specialist media (e.g. Hoyle's) and identified by biochemical tests and confirmed by 16s rRNA sequencing. The toxin gene is detected by nucleic acid amplification test (NAAT).

Prevention and control

Diphtheria is prevented by childhood vaccination with a toxoid (see Chapter 11). Immunity is long-lasting but boosters for adults at extra risk may be required (e.g. laboratory staff). Contacts of cases must be identified and given antibiotic prophylaxis, vaccination and/or specific antitoxin.

Tetanus
Epidemiology and pathogenesis

Infection occurs in wounds that are deep enough to produce anaerobic conditions. *Clostridium tetani* produces tetanospasmin, which prevents release of the inhibitory transmitter γ-aminobutyric acid (GABA), thereby resulting in muscle spasms. Neonatal tetanus,

Medical Microbiology and Infection at a Glance, Fourth Edition. Stephen H. Gillespie, Kathleen B. Bamford. © 2012 John Wiley & Sons, Ltd.
Published 2012 by John Wiley & Sons, Ltd.

which may occur if the umbilical stump is contaminated after delivery, is an important cause of death in developing countries. Tetanus is now rare in developed countries, usually being found in the elderly in whom immunity has declined. The disease may follow a trivial gardening injury.

Clinical features

Spastic paralysis and muscle spasms may develop at the site of the lesion and if untreated become generalized. Perioral muscle spasm (risus sardonicus) and spinal spasm (opisthotonus) may develop. Spasms are painful, may be stimulated by light or sudden noise and may compromise respiration so that secondary bacterial pneumonia may develop. Diagnosis is based on history and clinical features; isolation of the organism is not diagnostic.

Treatment and prevention

Treatment is with muscle relaxants and the use of human tetanus hyperimmune immunoglobulin and antibiotics to limit further toxin activity. Ventilation and treatment of secondary pneumonia may be required.

Infants are protected by passive immunity if their mothers are vaccinated. The disease is prevented by childhood immunization and boosters are given at school entry and every 10–15 years. Unvaccinated patients with tetanus-prone wounds should receive antibiotics and human tetanus immunoglobulin, followed by a course of vaccination.

Bordetella spp.

Bordetella pertussis and *B. parapertussis* can cause whooping cough. In the absence of an adequate vaccination campaign, epidemics of whooping cough occur in children every 4 years. Asymptomatic or unrecognized infection in adolescents and young adults maintains the cycle of infection in the human population.

Pathogenesis

Bordetella pertussis express fimbriae that aid their adhesion and produce a number of exotoxins that include pertussis toxin, adenyl cyclase and tracheal cytotoxin. There is a complex interaction with the cells of the respiratory tract that produces thickened bronchial secretions and paroxysmal cough. Complications include:
- secondary respiratory tract infection;
- apnoea following coughing spasms;
- raised intracranial pressure.

Clinical features

A 2-week, cold-like illness occurs before the characteristic cough is heard – repeated, prolonged coughing fits followed by an inspiratory whoop that may be absent in very young children and adults. Coughing may be associated with vomiting and subconjunctival haemorrhage; this phase can last for up to 3 months. Infection can be complicated by secondary pneumonia and otitis media.

Laboratory diagnosis

Specimens for culture are obtained using a pernasal swab, but the organism is difficult to isolate and NAATs are more likely to achieve a diagnosis.

Treatment

Erythromycin is thought to decrease infectivity and shorten symptoms if given early during the catarrhal phase. Symptomatic support and early treatment of secondary infections is the mainstay of treatment.

Prevention and control

An acellular vaccine is given as part of the childhood vaccination scheme.

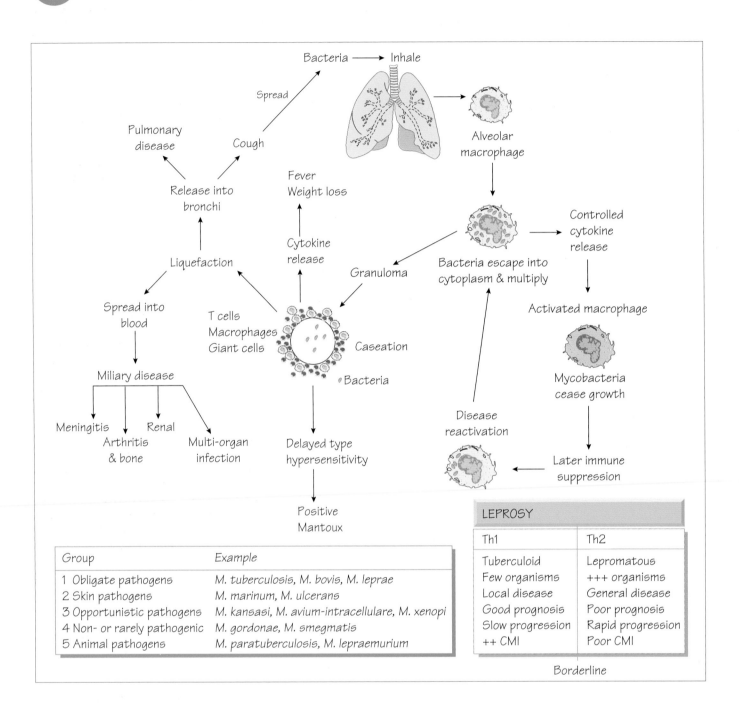

Group	Example
1 Obligate pathogens	M. tuberculosis, M. bovis, M. leprae
2 Skin pathogens	M. marinum, M. ulcerans
3 Opportunistic pathogens	M. kansasi, M. avium-intracellulare, M. xenopi
4 Non- or rarely pathogenic	M. gordonae, M. smegmatis
5 Animal pathogens	M. paratuberculosis, M. lepraemurium

LEPROSY

Th1	Th2
Tuberculoid	Lepromatous
Few organisms	+++ organisms
Local disease	General disease
Good prognosis	Poor prognosis
Slow progression	Rapid progression
++ CMI	Poor CMI

Borderline

Medical Microbiology and Infection at a Glance, Fourth Edition. Stephen H. Gillespie, Kathleen B. Bamford. © 2012 John Wiley & Sons, Ltd.
Published 2012 by John Wiley & Sons, Ltd.

Mycobacteria possess a lipid-rich cell wall that retains some dyes, resisting decolourization with acid (hence they are known as **acid-fast**). There are more than 50 species; although most are environmental organisms that rarely cause human infection.

Mycobacterium tuberculosis
Epidemiology and pathogenesis
Tuberculosis is spread from person to person by the airborne route. In childhood, the lung is the first site of infection and most infections resolve with local scarring (the primary complex). In about 5% the infection is not controlled and it spreads from the primary focus throughout the body (miliary spread). This may resolve spontaneously or develop into other localized infections (e.g. meningitis, osteomyelitis). Disease may emerge if immunity wanes or in later life (10% lifetime risk). Immunity depends on effective T-cell function so if compromised, individuals are more likely to develop symptomatic disease.

Mycobacterium tuberculosis is ingested by macrophages but escapes from the phagolysosome to multiply in the cytoplasm. The intense immune response causes local tissue destruction (cavitation in the lung) and cytokine-mediated systemic effects (fever and weight loss). Many antigens have been identified as possible virulence determinants, including lipoarabinomannan (stimulates cytokines) and superoxide dismutase (promotes intramacrophage survival).

Clinical features
Mycobacterium tuberculosis may affect every organ of the body mimicking both inflammatory and malignant diseases.
• **Pulmonary**: chronic cough, haemoptysis, fever and weight loss, or recurrent bacterial pneumonia. If untreated, it follows a chronic, deteriorating course.
• **Meningitis**: fever and deteriorating level of consciousness.
• **Renal**: signs of local infection, fever and weight loss, complicated by ureteric fibrosis and hydronephrosis.
• **Bone**: the lumbosacral spine is a common site of infection. Complicated by vertebral collapse and nerve compression. Pus may spread under the psoas sheath to appear in the groin (psoas abscess).
• **Joints**: infects large joints causing destructive arthritis.
• **Abdominal**: mesenteric lymphadenopathy and chronic peritonitis may present as fever, weight loss, ascites and intestinal malabsorption.
• **Miliary disease**: may occur without evidence of active lung infection.

Laboratory diagnosis
A wide range of specimens can be examined, using a variety of techniques.
• Direct staining with Ziehl–Neelsen's method or auramine.
• Culture on lipid-rich (egg-containing) medium with malachite green (Löwenstein–Jensen or L–J medium) to suppress other organisms.
• Automated growth detection, which can speed isolation.

• Susceptibility is usually tested in automated systems (e.g. Mycobacterial Growth Indicator Test [MGIT]).
• Nucleic acid amplification tests (NAATs) that include detection of drug resistance mutations allow rapid identification of patients with multidrug resistance TB (MDRTB) so they can be isolated and appropriately treated.
• Typing by identification of the number of copies of 23 repetitive DNA elements.
• Immunity is determined by measuring the γ-interferon response to specific antigens (Interferon Gamma Release Assay [IGRA]).

Treatment and prevention
The standard regimen for pulmonary infection is rifampicin and isoniazid for 6 months, with ethambutol and pyrazinamide for the first 2 months. Regimens for other sites are similar, taking into account drug penetration (e.g. into CSF). There is a rising trend of MDRTB. Features that predispose a patient to this include:
• previous incomplete treatment;
• being a known contact of an MDR patient;
• residence in a country where MDRTB is common;
• failure to respond to an adequate regimen.

Treatment of MDRTB is complex, requiring a combination of second-line agents, such as aminoglycosides, fluoroquinolones, ethionamide or cycloserine, guided by susceptibility tests.

Vaccination with attenuated bacille Calmette–Guérin (BCG) strain may protect against miliary spread, but trials in some countries have shown no benefit. Patients at high risk of developing tuberculosis may be given prophylaxis with isoniazid and rifampicin.

Mycobacterium leprae
Mycobacterium leprae attacks peripheral nerves, causing anaesthesia. Digital destruction and deformity follow, leaving the patient severely disabled. The end result of infection depends on the individual immune response, forming a spectrum from 'tuberculoid' dominated by a Th1 response, through 'borderline', to 'lepromatous' dominated by a Th2 response. Patients with tuberculoid disease have a stronger immune response, more localized disease and fewer bacteria. With lepromatous disease there is poor cell-mediated immunity (CMI) and generalized disease (leonine facies, depigmentation and anaesthesia).

Diagnosis is by Ziehl–Neelsen's stain of a split-skin smear and histological examination of a skin biopsy. Treatment with rifampicin, dapsone and clofazimine rapidly renders the patient non-infectious, but cannot alter nerve damage and deformity, which must be managed by remedial surgery.

Mycobacterium marinum and *Mycobacterium ulcerans*
Mycobacterium marinum causes a chronic granulomatous infection of the skin and is acquired from rivers, poorly maintained swimming pools or fish tanks. It is characterized by encrusted pustular lesions. *M. ulcerans* infection is associated with farming in Africa and Australia. The lower limb is usually affected with a papular lesion, which ulcerates and may destroy underlying tissue including bone.

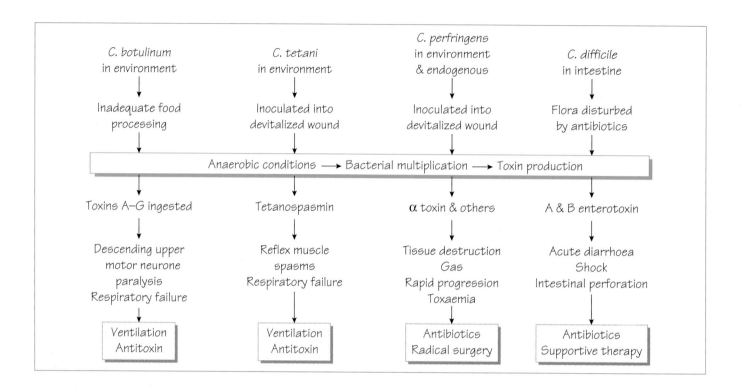

Clostridium spp. are anaerobic spore-forming organisms that are able to survive well in the environment. Their normal habitats are soil, water and the intestinal tract of humans and animals. Only a few of the 80 species are human pathogens. Toxin production is the main pathogenicity mechanism.

Clostridium difficile

Pseudomembranous colitis, an acute inflammatory diarrhoeal disease, is an important cause of morbidity and mortality in hospitals.

Epidemiology and pathogenesis

The frequency of *Clostridium difficile* carriage increases with duration of hospital stay. When the gut microflora is disturbed by antibiotics overgrowth can occur. Enterotoxins A and B and binary toxin production cause tissue damage and fluid diarrhoea. Some strains that are fluoroquinolone resistant and have evidence of enhanced toxin production are associated with more severe disease and extensive hospital outbreaks.

Clinical features

- History of previous antibiotic exposure.
- Three or more loose/unformed stools per day.
- Possible development of abdominal pain.
- Pseudomembranes seen on sigmoidoscopy on the mucosal surface of the rectum and sigmoid colon.
- Possible complications of toxic megacolon, bowel perforation and systemic toxicity, which are associated with high mortality.

Diagnosis

- Detection of toxin or glutamate dehydrogenase (GDH) by enzyme immunoassay (EIA).
- Detection of toxin genes by nucleic acid amplification test (NAAT).
- Typing – generally by ribotyping, but multilocus variable-number tandem repeat analysis is superseding this.

Treatment and prevention

- Stop the inciting (antibiotic) agent.
- Treat with oral metronidazole for 10 days.
- Oral vancomycin plus intravenous metronidazole for severe cases and treatment failures.
- Relapse occurs in up to 25% of patients.
- Rapid and strict isolation is essential.
- Enhanced ward cleaning and attention to hand hygiene are essential (Awareness may be raised by 'SIGHT' – Suspect Isolate Gloves and aprons Hand hygiene Toxin test).

Botulism

Clostridium botulinum is found in the environment and can contaminate wounds and food. Of the seven serotypes, A, B and E are most commonly associated with human disease.

Epidemiology and pathogenesis

Incomplete heat treatment of contaminated food in canning or bottling processes allows this organism to survive and produce toxin. Botulinum toxin, a neurotoxin, inhibits the release of neu-

Medical Microbiology and Infection at a Glance, Fourth Edition. Stephen H. Gillespie, Kathleen B. Bamford. © 2012 John Wiley & Sons, Ltd.

rotransmitters. Clinically, there are three forms of the disease: food intoxication, wound botulism and infant botulism.

Clinical features
• Rapid onset – within 6h of ingestion.
• Descending flaccid paralysis, beginning with the cranial nerves.
• Dysphagia and blurred vision, followed by more general paralysis.
• Sensory function is normal.
• Infants appear floppy and listless, constipated and have generalized muscle weakness.

Diagnosis is based on the clinical features and a history of ingestion of suspect food. Toxin may be detected in food, faeces and serum by EIA. Botulinum neurotoxin genes can be detected by NAAT.

Treatment and prevention
Treatment is with specific antitoxin and ventilatory support. Ventilator-associated pneumonia is an important complication. The disease is prevented by adequate process control in the food-processing industry and home preservation.

Gas gangrene
Clostridium perfringens is the organism most commonly associated with gas gangrene, but *C. septicum*, *C. novyi*, *C. histolyticum* and *C. sordellii* can also be implicated. *C. perfringens* is capsulate and produces a range of toxins, of which lecithinase C (α-toxin) is the most important.

Epidemiology and pathogenesis
• At-risk wounds are those where devitalized tissue is contaminated by environmental spores.

• Spores germinate and organisms multiply in the ischaemic conditions.
• Toxin is released.
• Toxin-mediated tissue damage creates more anaerobic conditions and progression is rapid.
• Infections can be mixed with other organisms especially in drug users.

Clinical features
• Onset within 3 days of injury.
• Pain in the wound.
• Tenseness of the skin, which develops an underlying blue discoloration, foul smell and crepitus.
• Toxaemia producing circulatory shock.

Diagnosis
The diagnosis is made clinically and treatment must not wait for laboratory confirmation, which is rarely available.

Treatment and prevention
Treatment depends on debridement of devitalized tissue and intravenous antibiotics. Hyperbaric oxygen may also be beneficial. The condition is prevented by good management of potentially infected, devitalized wounds.

Clostridium perfringens food poisoning
This condition is typically associated with meals containing meat that are cooled slowly and subsequently reheated. Surviving clostridia release toxin in the stomach when they form spores; this leads to nausea, vomiting and diarrhoea. An EIA to detect toxin in faeces is available. Rarely, gut infection with clostridia causes severe enteritis.

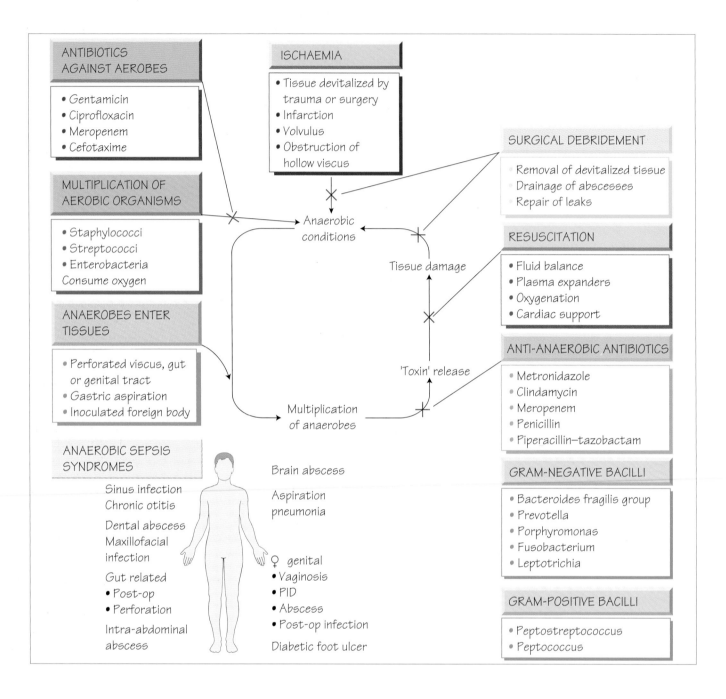

Non-sporing anaerobes form the major part of the normal human bacterial flora, outnumbering all other organisms in the gut by a factor of 10^3. They are also found in the genital tract, oropharynx and skin.

Anaerobic sepsis
Pathogenesis
- Infection with non-sporing anaerobes is usually endogenous.
- Normal flora may escape into a sterile site following perforation of a hollow viscus (e.g. large intestine).

- Ischaemia (e.g. strangulated hernia) permits anaerobic growth, or the metabolism of facultative bacteria produce anaerobic conditions (e.g. in deep skin ulcers or intraperitoneal infection).
- Once established, anaerobic multiplication is promoted by release of toxic metabolic products and proteolytic enzymes.
- Toxic products of inflammatory cells, such as reactive oxygen intermediates, exacerbate tissue damage creating a vicious cycle of anaerobic sepsis that is rapidly progressive.

Medical Microbiology and Infection at a Glance, Fourth Edition. Stephen H. Gillespie, Kathleen B. Bamford. © 2012 John Wiley & Sons, Ltd.

Clinical syndromes due to non-sporing anaerobes

• Intra-abdominal sepsis may follow spontaneous bowel perforation or postsurgical leakage and may lead to abscess formation (e.g. abdominal or liver abscesses).

• Sepsis of the female genital tract is often secondary to septic abortion, prolonged rupture of the membranes, complicated caesarean section or retained products of conception.

• Non-sporing anaerobes are implicated in pelvic inflammatory disease (PID).

• Imbalance in the anaerobic flora of the vagina may lead to the non-specific vaginosis syndrome (see Chapter 51).

• Non-sporing anaerobes play a part in polymicrobial liver abscesses and biliary sepsis.

• Pneumonia following aspiration or associated with carcinoma or foreign-body obstruction has a significant anaerobic component and lung abscesses can develop.

• Brain abscesses often have an important anaerobic component.

• Chronic paranasal suppuration, such as in chronic otitis media and chronic sinusitis, may contain non-sporing anaerobes.

• Anaerobes may colonize chronic skin ulcers.

• Rarely tropical ulcers may be caused by *Fusobacterium ulcerans*.

Laboratory diagnosis

Non-sporing anaerobes are nutritionally fastidious, sensitive to oxygen and difficult or slow to grow. Specimens should be plated directly in theatre or at the bedside, or transported rapidly to the laboratory in an anaerobic transport system. Pus rather than swabs (which dry out quickly) should be sent. There is an increasing role for nucleic acid amplification tests (NAATs) based on amplification of the 16S rRNA gene.

Anaerobic species are identified by phenotypic laboratory tests, by studying the end products of metabolism using gas–liquid chromatography and, increasingly, by molecular means.

Antibiotic susceptibility

Almost all anaerobes are susceptible to metronidazole, although resistance has been reported. Other active agents include meropenem, piperacillin–tazobactam, clindamycin, chloramphenicol, penicillin and erythromycin. (**Note**: Penicillin and erythromycin are not active against *Bacteroides fragilis*, the anaerobe most commonly isolated from abdominal sepsis.)

Management

Effective management depends on surgery and antimicrobial therapy. Surgical procedures include:

• drainage of abscesses;

• closure of perforations;

• resection of gangrenous tissue;

• debridement of non-viable tissue from ulcers;

• treatment of coexisting infection.

Metronidazole is the most commonly used anti-anaerobic agent.

Prevention and control

The risk of anaerobic infection in elective surgery can be reduced by good operative technique and perioperative antibiotics with anti-anaerobic activity (see Chapters 5 and 6).

Pathogens of anaerobic sepsis

Bacteroides fragilis is typically associated with postoperative sepsis in abdominal and gynaecological surgery. It also contributes to the polymicrobial flora found in cerebral, hepatic and lung abscesses.

• The most common agent of serious anaerobic sepsis.

• Penicillin resistant through β-lactamase production.

• Produces a protease (DNAse), heparinase and neuraminidase.

• Has an antiphagocytic capsule.

Prevotella melaninogenicus and *Fusobacteria* are found chiefly in the oral cavity. They are associated with periodontal disease, gingivitis, dental abscess, sinus infection, cerebral and lung abscesses, and necrotizing pneumonia. They are also found in association with *Borrelia vincentii* in Vincent's angina, and in ulcerative diseases such as cancrum oris (Ludwig's angina), both of which affect the head and neck. They may contribute to anaerobic cellulitis.

Peptococcus and *Peptostreptococcus* are the only anaerobic Gram-positive cocci that are regularly found in human specimens. They are often found in mixed infections, such as dental sepsis, cerebral or lung abscesses, and soft-tissue and wound infections. They are also associated with necrotizing fasciitis, where a mixed infection of anaerobic cocci, facultative streptococci and sometimes *Staphylococcus aureus* usually progresses rapidly, with destruction of the skin and deeper tissues that leads to septicaemia and death.

Actinomyces spp. are associated with chronic abscesses following dental sepsis, lung abscesses, gut perforation and infection of intrauterine devices. Long-term penicillin is usually effective.

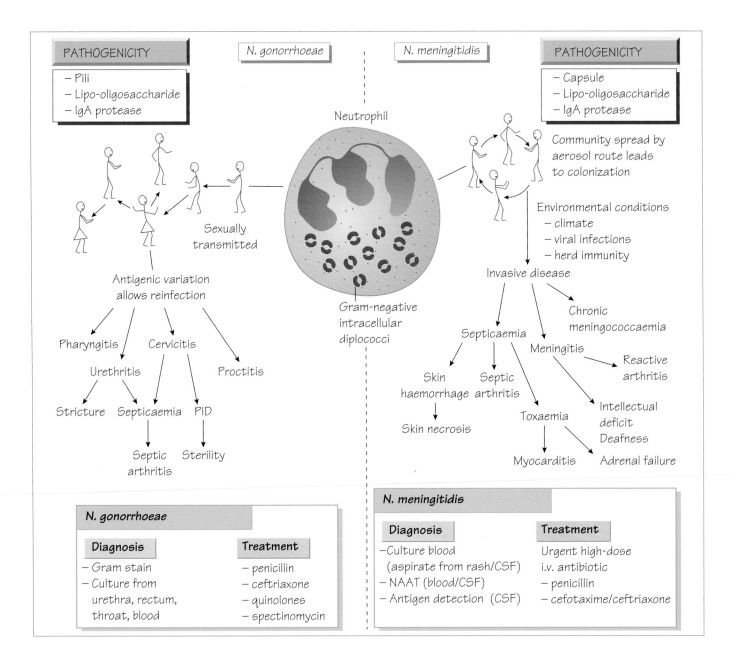

PATHOGENICITY

– Pili
– Lipo-oligosaccharide
– IgA protease

N. gonorrhoeae

N. meningitidis

PATHOGENICITY

– Capsule
– Lipo-oligosaccharide
– IgA protease

Neutrophil

Sexually transmitted

Community spread by aerosol route leads to colonization

Environmental conditions
– climate
– viral infections
– herd immunity

Invasive disease

Antigenic variation allows reinfection

Gram-negative intracellular diplococci

Chronic meningococcaemia

Septicaemia

Meningitis

Reactive arthritis

Pharyngitis / Cervicitis

Urethritis / Proctitis

Skin haemorrhage

Septic arthritis

Stricture Septicaemia PID

Skin necrosis

Toxaemia

Intellectual deficit
Deafness

Septic arthritis Sterility

Myocarditis

Adrenal failure

N. gonorrhoeae

Diagnosis	Treatment
– Gram stain – Culture from urethra, rectum, throat, blood	– penicillin – ceftriaxone – quinolones – spectinomycin

N. meningitidis

Diagnosis	Treatment
–Culture blood (aspirate from rash/CSF) – NAAT (blood/CSF) – Antigen detection (CSF)	Urgent high-dose i.v. antibiotic – penicillin – cefotaxime/ceftriaxone

Neisseria gonorrhoeae

Infection with *Neisseria gonorrhoeae*, a Gram-negative diplococcus, is most common in individuals between 15 and 35 years of age. It is almost exclusively spread by sexual contact.

Pathogenesis

The organism adheres to the genitourinary epithelium via pili, then invades the epithelial layer and provokes a local acute inflammatory response. Variation in the proteins of the pili means that infection does not provide protection against re-infection, therefore infections with another strain of different antigenic structure are possible.

Clinical features

• Acute painful urethritis and urethral discharge.
• Female infection (cervicitis) is often asymptomatic or associated with vaginal discharge.
• Pelvic inflammatory disease (PID) may develop (see Chapter 51).
• Pharyngeal infection causes sore throat.
• Rectal infection causes a purulent proctitis.
• Infection can be complicated by bacteraemia, septic or reactive arthritis of the large joints, or pustular rashes.
• Late complications include female infertility and male urethral stricture.

Medical Microbiology and Infection at a Glance, Fourth Edition. Stephen H. Gillespie, Kathleen B. Bamford. © 2012 John Wiley & Sons, Ltd.

Diagnosis

The optimal diagnostic technique is a nucleic acid amplification test (NAAT) on urethral or vaginal swabs, or urine. Positive samples are then cultured for susceptibility testing.

Treatment and prevention

Treatment must be given before susceptibility results are available and is based on the known susceptibility patterns found at the clinic, as emergence of resistance is a problem. Ceftriaxone, spectinomycin or fluoroquinolones may be used. Gonorrhoea can be prevented by avoiding sexual contact with individuals at high risk and using effective barrier contraception. Contacts of infected individuals should be traced and treated. At present vaccine development is precluded by the antigenic variation that occurs within the pili.

Neisseria meningitidis
Epidemiology

Carriage of *Neisseria meningitidis* (meningococcus) is common; actual disease only develops in a few individuals. Infection is most common in the winter, with epidemics occurring every 10–12 years. In Africa, severe epidemics of group A infection occur in the 'meningitis belt' where the incidence can rise to 1000 cases per 100000 each year. Most invasive infections are caused by serogroups A, B or C. Group B infection is now the commonest, as the incidence of group C infection has reduced in communities where vaccination has become routine. A group A vaccination programme is being implemented in Africa.

Pathogenesis and clinical features

• An antiphagocytic polysaccharide capsule that allows survival in the bloodstream.
• Lipo-oligosaccharide activates complement and stimulates cytokine release, which leads to shock and disseminated intravascular coagulation (DIC).

• The organism hijacks the β2-adrenoreceptor to cross the brain vascular endothelium.
• Meningococci cross mucosal epithelium by endocytosis.

Meningococcal meningitis is characterized by fever, neck stiffness and reduced consciousness. The petechial rash, a sign of septicaemia, may be present without other signs of meningitis. Septic or reactive arthritis may develop.

Diagnosis and treatment

The diagnosis is usually made clinically and confirmed by culture of blood, aspirate from the rash and CSF. Rapid antigen detection or NAAT on CSF and blood are sensitive and reliable.

Infection is life-threatening and rapidly progressive; treatment should not await laboratory confirmation or hospitalization. Intravenous benzylpenicillin (intramuscular in the community setting) is the antibiotic of choice, but there have been reports of meningococci with reduced susceptibility in other countries and cefotaxime is an alternative. Treatment does not eradicate carriage so the patient should be given 'prophylaxis' following recovery.

Prevention

• A protein-conjugated serogroup C vaccine is more than 90% efficient and has vastly reduced the incidence where it has been introduced.
• An effective vaccine against serogroup B is not available, although a vaccine based on membrane proteins specific for epidemic strains has shown promise.
• Close contacts of patients with meningococcal meningitis should be given 'prophylaxis' with rifampicin or ciprofloxacin.

Moraxella catarrhalis

This Gram-negative coccobacillus is usually a commensal of the upper respiratory tract. It is associated with otitis media, sinusitis and lower respiratory tract infection in children or patients with chronic pulmonary disease. It usually produces β lactamase

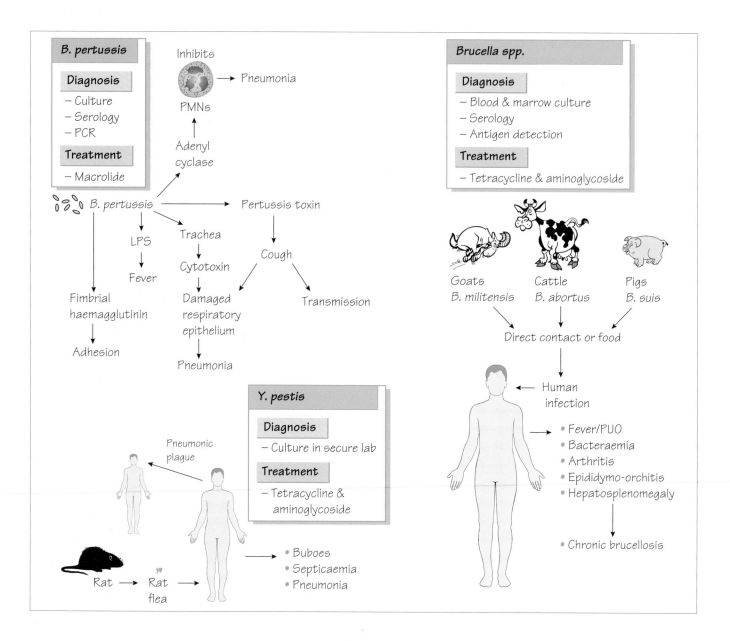

Haemophilus

Haemophilus spp. are fastidious Gram-negative coccobacilli that colonize mucosal surfaces. *H. influenzae* and *H. ducreyi* are the main pathogenic species.

Haemophilus influenzae

Haemophilus influenzae expresses an antiphagocytic polysaccharide capsule of which there are six types (a–f). It also expresses a lipopolysaccharide (LPS) and an IgA1 protease. Septicaemia, meningitis and osteomyelitis are usually associated with type b infection in individuals who have not been vaccinated.

Clinical features

Infection occurs in preschool children, causing pyogenic meningitis, acute epiglottitis, septicaemia, facial cellulitis or osteomyelitis. Non-capsulate strains are usually commensal in the nasopharynx, but may cause adult otitis media, sinusitis, and chest infection in patients with obstructive airways disease.

Laboratory diagnosis

Antigen detection provides rapid diagnosis in meningitis. Culture of CSF, sputum, blood or pus is used. Increasingly, *H. influenzae* is diagnosed as part of multiplex nucleic acid amplification tests (NAATs).

Medical Microbiology and Infection at a Glance, Fourth Edition. Stephen H. Gillespie, Kathleen B. Bamford. © 2012 John Wiley & Sons, Ltd.
Published 2012 by John Wiley & Sons, Ltd.

Treatment and prevention

Many *H. influenzae* express a β-lactamase and are ampicillin resistant. Co-amoxiclav, clarithromycin, tetracycline or trimethoprim can be used. Severe infections are treated with a β-lactam-stable cephalosporin.

A protein-conjugated polysaccharide vaccine against type b has almost eradicated childhood infection. Non-capsulate *Haemophilus* is ubiquitous and predisposed patients cannot avoid infection.

Haemophilus ducreyi

Haemophilus ducreyi is transmitted sexually and causes painful, irregular, soft genital ulcers (chancroid). There is associated lymphadenopathy, and suppurating inguinal lymph nodes may lead to sinus formation. Infection is more common in developing countries and facilitates the transmission of HIV.

Transmission is controlled by treatment with azithromycin, ceftriaxone or a fluoroquinolone coupled with efficient contact tracing (see Chapter 8).

Brucella spp.

Brucella melitensis, *B. abortus* and *B. suis* have goats, cattle and pigs, respectively, as their main hosts. They are aerobic or capnophilic and require serum-containing medium to grow.

Brucella infection spreads to humans through direct contact with domesticated animals or their products (e.g. unpasteurized milk). Vets, farmers and abattoir workers are at increased risk of infection.

Pathogenesis

Brucellae are able to survive inside the cells of the reticuloendothelial system using superoxide dismutase and nucleotide-like substances to inhibit the intracellular killing mechanisms of their host.

Clinical features

• Intermittent, high fever in the early stages of infection, giving rise to its old name 'undulant fever'.
• Myalgia, arthralgia and lumbosacral tenderness.
• Complications of acute infection that include septic arthritis, osteomyelitis and epididymo-orchitis.
• A chronic infection that may develop without treatment, which may resolve or continue to give symptoms, often accompanied by psychiatric complaints, for many years.

Laboratory diagnosis

Culture of blood and bone marrow is diagnostic, although culture is less likely to be positive in chronic disease. Incubation, in a high containment facility, must be continued for up to 3 weeks. Serodiagnosis is by enzyme immunoassay (EIA) to detect both IgG and IgM.

Treatment and prevention

Optimal treatment is with tetracycline for 1 month. Streptomycin should be added for patients with complications. Transmission by food can be prevented by pasteurization. Appropriate animal husbandry techniques can reduce the risk of occupational infection. An animal vaccine is available but is not sufficiently safe for human use. Animal control measures have eradicated brucellosis from farms in many countries.

Francisella tularensis

This pathogen of rodents and deer can be found in North America and northern Europe. Infection, which is spread by the aerial route, direct contact with wild animals or by tick bite, is rare, being found mainly in campers and hunters.

Infection may be ocular or localized to the skin, with regional lymphadenopathy. Systemic infection gives a syndrome that resembles typhoid, with 5–10% mortality. Diagnosis is by serology or by culture. Treatment is with tetracycline.

Yersinia
Yersinia pestis

Yersinia pestis infection is described in Chapter 54.

Yersinia enterocolitica

This organism, morphologically and biochemically similar to *Y. pestis*, causes acute enteritis, mesenteric adenitis and, rarely, septicaemia. It is transmitted to humans in food and water. Infection may be complicated by polyarthritis and erythema nodosum. Patients with iron-overload syndromes are especially susceptible. Diagnosis is by isolation from faeces, blood or lymph node, or detection of antibodies. Treatment with ciprofloxacin or co-trimoxazole is indicated in serious infection; tetracycline is an alternative therapy.

Yersinia pseudotuberculosis

This organism can cause mesenteric adenitis that mimics appendicitis.

Bartonella spp.

Bartonella are small Gram-negative bacteria that can invade host red blood cells, epithelial and bone marrow cells. *B. henselae* is responsible for cat-scratch disease (see Chapter 54), endocarditis and bacilliary angiomatosis, a febrile illness associated with a red, papular rash that is commonly found in patients with AIDS. *B. quintana,* which is transmitted by lice, causes a relapsing febrile illness found in severely disadvantaged individuals. *B. bacilliformis* infection can cause Oroya fever, an acute febrile haemolytic anaemia or mild fever with body pain, nausea and headache. It is transmitted by sandflies and found only in Peru and neighbouring countries. Diagnosis is by culture, but NAATs and sequencing are more sensitive.

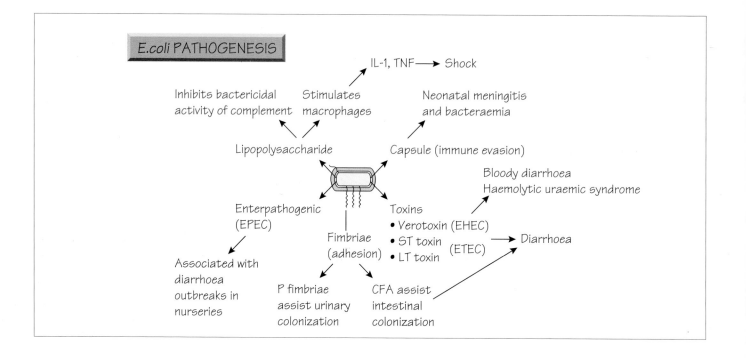

The Enterobacteriaceae are a large family (>20 genera and 100 species) of facultatively anaerobic Gram-negative bacilli that are easily cultured, reduce nitrate and ferment glucose. The wide diversity of named species is in part because they are easy to grow and study in the laboratory, but also because they are capable of causing a wide variety of clinical syndromes.

Habitat and transmission

The Enterobacteriaceae are almost ubiquitous organisms. They are found in:
• the normal flora of animals and humans (forming a major component);
• water;
• soil.

Transmission is both from other animals or humans and from the inanimate environment. Many infections arise from the body's normal flora when opportunities are provided by medical, surgical or other therapy (see Chapter 2).

Pathogenicity
Capsules

Many species produce extracellular capsular polysaccharides (e.g. *Klebsiella* spp., *Escherichia coli* and *Salmonella typhi*). *S. typhi* possesses a capsule or Vi (virulence) antigen and a vaccine containing the Vi antigen is protective against typhoid. *E. coli* K1 is the most common type of *E. coli* isolated from patients with neonatal meningitis and septicaemia. There are biochemical and structural similarities between *E. coli* K1 and *N. meningitidis* group B and

human central nervous system antigens that may give these pathogens an advantage.

Lipopolysaccharide

The lipopolysaccharide (LPS) molecule consists of a central lipid A and oligosaccharide core, and a long straight or branched polysaccharide 'O' antigen. It is located in the bacterial outer membrane and protects the organism against the bactericidal activity of complement. Lipid A stimulates host macrophages to produce cytokines, such as interleukin-1 and tumour necrosis factor (TNF), which mediate the fever, shock and metabolic acidosis associated with severe sepsis. Some clinical syndromes are associated with particular O antigens: for example, *E. coli* O157 may produce verotoxin, which causes haemolytic uraemic syndrome (HUS); other O types are associated with urinary tract infection or diarrhoea. However, these are merely temporal relationships between a variety of bacterial characteristics that include an O antigen and a particular virulence determinant.

Urease

Proteus spp. express a potent urease. In the urinary tract, urea lowers the pH, therefore infection with *Proteus* spp. allows calcium and phosphates to precipitate, with the formation of renal stones (see Chapter 51).

Fimbriae

Fimbriae or pili are bacterial organelles that promote colonization, for example in the ureter. *E. coli* that express mannose-binding fimbriae are associated with lower urinary tract infections and cystitis, whereas those that express P fimbriae are associated with pyelonephritis and septicaemia. In the intestine, *E. coli* that express different fimbriae (colonization factor antigens, [CFAs]) have been associated with diarrhoea.

Toxins

Enterotoxigenic *E. coli*

Enterotoxigenic *E. coli* (ETEC) produce LT and ST toxins that act on the enterocyte to stimulate fluid secretion, resulting in diarrhoea. LT toxin, which is heat labile, shares 70% homology with cholera toxin and, like cholera toxin, increases local cyclic adenosine monophosphate (cAMP) in the enteric cell. ST toxin is heat stable and stimulates cyclic guanyl monophosphate (cGMP). *E. coli* that possess these enterotoxins are associated with travellers' diarrhoea, which is a short-lived, watery diarrhoeal disease.

Enteroaggregative *E. coli*

Some strains of *E. coli*, which are known as enteroaggregative *E. coli* (EAggEC), secrete plasmid-encoded toxin, a serine protease that binds α-fodrin and causes disruption of the actin cytoskeleton, and are able to cause chronic diarrhoea. Strains express an ST-like toxin or a haemolysin-like toxin.

Enteropathogenic *E. coli*

Enteropathogenic *E. coli* (EPEC) cause disease by colonizing the epithelial lining of the small intestine and injecting effector proteins that cause effacement of microvilli and intimate adherence. Isolates with this characteristic were the first *E. coli* recognized as primary pathogens, when they caused outbreaks of diarrhoea in preschool nurseries.

Enterohaemorrhagic *E. coli*

The enterohaemorrhagic *E. coli* (EHEC) strains produce a verotoxin named because of its *in vitro* activity on 'vero' cells. The haemorrhagic diarrhoea that they cause can be complicated by haemolysis and acute renal failure (HUS). This organism is commensal in cattle and is transmitted to humans through hygiene failure in abattoirs and food production. A similar toxin (Shiga toxin) is a major virulence determinant in *Shigella dysenteriae*.

Genetic exchange

The Enterobacteriaceae can gain DNA rapidly from other organisms through transposons, integrons or plasmids. This enables antibiotic-resistance genes to spread from one species to another. In the hospital environment the survival of antibiotic-resistant strains is favoured. In some hospitals there have been outbreaks of multidrug-resistant *Klebsiella pneumoniae* in intensive care units. The Enterobacteriaceae have also been able to gain pathogenicity determinants by genetic exchange. Acquisition of a series of connected genes can occur and these are known as **pathogenicity islands**. In this way *Salmonella* have gained a series of genes that enable them to invade intestinal cells.

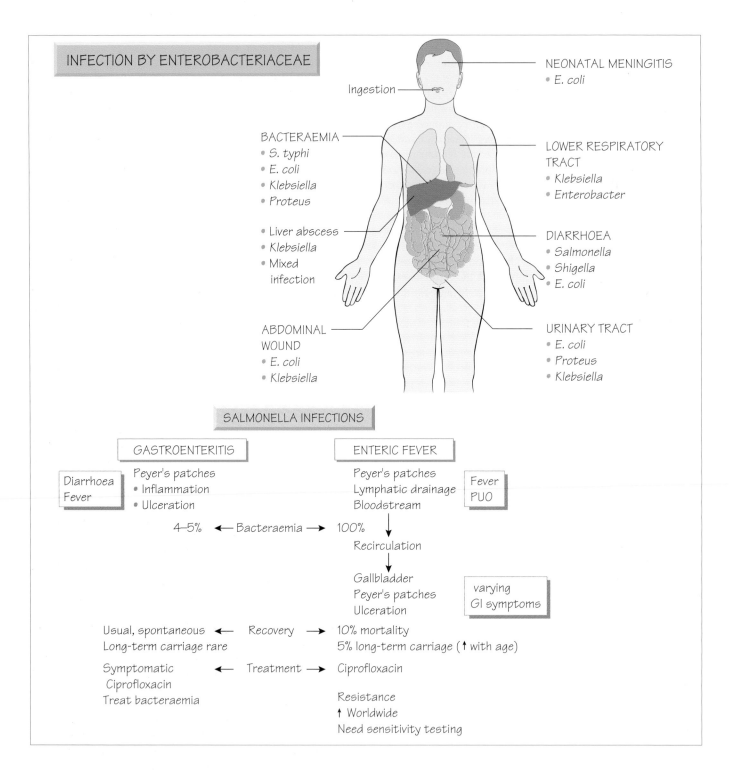

INFECTION BY ENTEROBACTERIACEAE

Ingestion

NEONATAL MENINGITIS
• E. coli

BACTERAEMIA
• S. typhi
• E. coli
• Klebsiella
• Proteus

• Liver abscess
• Klebsiella
• Mixed
  infection

LOWER RESPIRATORY
TRACT
• Klebsiella
• Enterobacter

DIARRHOEA
• Salmonella
• Shigella
• E. coli

ABDOMINAL
WOUND
• E. coli
• Klebsiella

URINARY TRACT
• E. coli
• Proteus
• Klebsiella

SALMONELLA INFECTIONS

GASTROENTERITIS

ENTERIC FEVER

Diarrhoea
Fever

Peyer's patches
• Inflammation
• Ulceration

Peyer's patches
Lymphatic drainage
Bloodstream

Fever
PUO

4–5% ← Bacteraemia → 100%

Recirculation

Gallbladder
Peyer's patches
Ulceration

varying
GI symptoms

Usual, spontaneous ← Recovery → 10% mortality
Long-term carriage rare   5% long-term carriage (↑ with age)

Symptomatic   ← Treatment → Ciprofloxacin
 Ciprofloxacin
Treat bacteraemia   Resistance
                                 ↑ Worldwide
                                 Need sensitivity testing

Medical Microbiology and Infection at a Glance, Fourth Edition. Stephen H. Gillespie, Kathleen B. Bamford. © 2012 John Wiley & Sons, Ltd.
Published 2012 by John Wiley & Sons, Ltd.

Salmonella

There are two *Salmonella* species that account for the majority of human and animal infections: *S. enterica* and *S. bongori*.

Salmonellosis

Salmonella are host-adapted to animals and humans. Infection is usually confined to the bowel, appearing as acute self-limiting diarrhoea. Less commonly invasive disease develops, which may be complicated by bacteraemia, life-threatening septicaemia or osteomyelitis.

Important sources of infection include domestic animals and the food produced from them, and human cases including convalescent carriers. Transmission is by the oral route, usually through ingestion of contaminated foods. Infection is commoner and often more severe in patients with reduced gastric acid and those who are immunocompromised or splenectomized. Reactive arthritis or a chronic carrier state may be complications of infection.

Enteric fever

Enteric fever (typhoid) is caused by *S. enterica* serotype Typhi or Paratyphi. Invasion of the intestinal wall, with spread to local lymph nodes, is followed by primary bacteraemia and infection of the reticuloendothelial system. The bacteria reinvade the bloodstream and gut from the gallbladder, multiply in Peyer's patches, and cause ulceration that may be complicated by haemorrhage or perforation. Patients present with fever, alteration of bowel habit (diarrhoea or constipation) and, more rarely, the classical rash (rose spots on the abdomen). Hepatosplenomegaly may also be demonstrated. Enteric fever may be complicated by osteomyelitis and, rarely, by meningitis.

Other infections

Urinary tract infection and pyelonephritis

Most *Escherichia coli* urinary tract infections are caused by a limited number of serotypes related to specialized adaptations (e.g. presence of the K antigen, adherence to uroepithelial cells via pili and haemolysin production). Mannose-resistant pili are associated with pyelonephritis. *Proteus* spp. also possess specialized adherence pili that mediate attachment to urinary epithelium. Urease production by *Proteus* is the most important virulence determinant in urinary infection, lowering pH and precipitating stone formation.

Meningitis and brain abscess

Escherichia coli is an important cause of neonatal meningitis associated with a high mortality. Strains often express copious amounts of K1 capsular antigen. Meningitis may also follow neurosurgical procedures, especially when prosthetic devices are inserted. Enterobacteriaceae are often found as part of the polymicrobial flora of brain abscess.

Osteomyelitis and septic arthritis

Salmonella osteomyelitis or septic arthritis is an important complication for patients with sickle cell disease or AIDS and for older patients. Infection with other Enterobacteriaceae can also follow penetrating trauma when contaminated fragments are taken into the bony tissue. Treatment often includes a fluoroquinolone such as ciprofloxacin that penetrates into bony tissue.

Klebsiella infections

Infection with *Klebsiella* spp. is usually acquired in a hospital environment. These organisms are an important cause of ventilator-associated pneumonia, urinary tract infection, wound infection and bacteraemia. Outbreaks of infection in high-dependency patients have been described and are associated with septicaemia and a high mortality. Primary pneumonia with *K. pneumoniae* subspecies *pneumoniae* is a rare, severe, community-acquired infection, associated with a poor outcome. *K. rhinoscleromatis* causes a progressive granulomatous infection of the nasal passages and surrounding mucous membranes. Most infections are found in the tropics. *K. ozanae* has been associated with chronic bronchiectasis.

Enterobacter, *Serratia* and *Citrobacter* infections

These are environmental organisms that may colonize and infect hospitalized patients, causing wound infections, bacteraemia and hospital-acquired pneumonia. Many isolates may be naturally resistant to antibiotics and treatment choices are limited.

Diagnosis

The Enterobacteriaciae grow readily on laboratory media and are identified by biochemical reactions, such as the pattern of fermentation of different sugars. Epidemiological investigation uses serotyping (sera directed against the lipopolysaccharide [O] antigens and flagellar [H] antigens), but molecular typing methods are gaining a place (e.g. sequencing of flagellar genes). In the case of typhoid, a diagnosis may be made by isolating organisms from the blood or bone marrow.

Treatment and prevention

Most enteric Gram-negative organisms are susceptible to aminoglycosides, extended-spectrum cephalosporins, fluoroquinolones, β-lactams and carbapenems (e.g. meropenem). As some produce β-lactamases and aminoglycoside-degrading enzymes, treatment should be guided by sensitivity tests. Emerging extended-spectrum β-lactamase-carrying strains (ESBLs) increase the resistance to broad-spectrum antibiotics.

In urinary tract infections, cefalexin, ampicillin, nitrofurantoin or trimethoprim are the first-choice antibiotics.

Diarrhoeal disease can be avoided by good hygiene, food preparation and safe water supplies. Treatment is primarily by oral rehydration (see Chapter 53).

Ceftriaxone and ciprofloxacin are used in the treatment of typhoid. Multidrug-resistant typhoid has been a major problem in some countries. A live attenuated vaccine (Ty21A) and a subcellular vaccine (containing the Vi antigen) are available for travellers to high-risk areas, but give only partial protection.

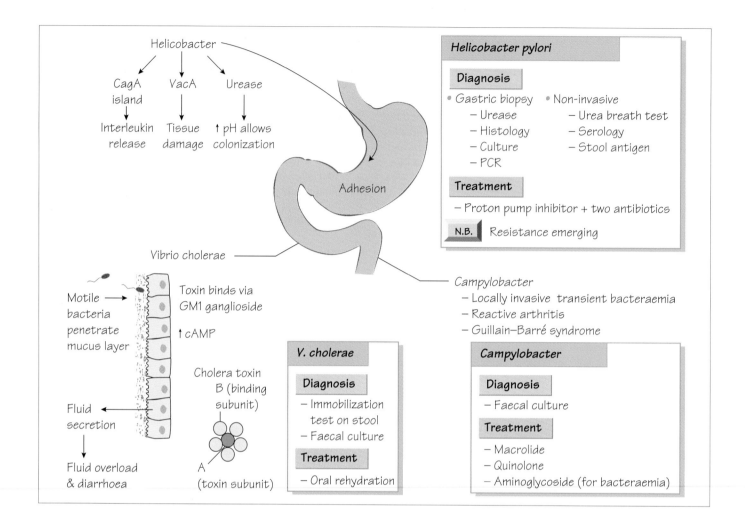

Vibrio spp.

Vibrios are Gram-negative, curved, motile bacilli. There are more than eight species; *V. cholerae* and *V. parahaemolyticus* being the main human pathogens.

Vibrio cholerae

The organism is subdivided by the somatic O antigens; O1 and O139 being the main types associated with cholera. They are able to survive the gastric acidity, burrowing through the intestinal mucus to attach to intestinal epithelial cells via the GM1 ganglioside and produce a multimeric protein toxin (cholera toxin), which stimulates adenyl cyclase within the enteric cells, resulting in the secretion of water and electrolytes into the lumen of the bowel.

Epidemiology

• Cholera is an exclusively human illness.
• It is transmitted via contaminated water and food.
• It is usually found in developing countries where there is inadequate sanitation and an unsafe water supply.

• Epidemics are facilitated by war, refugee movements and mass migration.
• Cholera gives rise to periodic pandemics: the most recent epidemic was the seventh recorded.

Clinical features

The clinical features of cholera include:
• massive, painless, fluid diarrhoea (up to 20 L per day);
• vomiting;
• severe dehydration and electrolyte imbalance.
   However, it can be a mild or asymptomatic condition.

Diagnosis

Where cholera is endemic, the diagnosis is based on clinical features. Immobilization of cholera bacteria in a diarrhoeal stool with specific antiserum yields a rapid diagnosis. The organism can be cultivated on specialist medium. Biochemical identification and serotyping should be performed to confirm the diagnosis.

Medical Microbiology and Infection at a Glance, Fourth Edition. Stephen H. Gillespie, Kathleen B. Bamford. © 2012 John Wiley & Sons, Ltd.

Treatment

- Oral rehydration solution (salt and glucose mix).
- Intravenous fluids for severe cases.
- Tetracycline or ciprofloxacin can shorten the duration and reduce the severity.

Prevention and control

- Safe water supplies are the mainstay of prevention.
- Community education reduces spread.
- Experimental live attenuated and subunit vaccines are under trial.

Campylobacter spp.

Campylobacter are associated with diarrhoeal disease and are the commonest cause of intestinal infection. Although there are more than 18 species of Campylobacter, C. jejuni is responsible for 90% of Campylobacter gastrointestinal infections. Infection follows ingestion of contaminated meat, poultry, unpasteurized milk or contaminated water. C. coli causes bacteraemia in immunocompromised patients.

Pathogenicity

Campylobacter jejuni invades and colonizes the mucosa of the small intestine. Antibodies to GM1 ganglioside are associated with Guillain–Barré syndrome.

Clinical features

- Influenza-like symptoms.
- Crampy abdominal pain.
- Diarrhoea, which may be blood stained.
- Children may be misdiagnosed as appendicitis or intussusception.
- Self-limiting bacteraemia is common.
- Guillain–Barré syndrome, which is ascending demyelination with motor and sensory deficits following a few weeks after Campylobacter infection, is a rare complication.
- Reactive arthritis may also occur.

Diagnosis

- Culture of faeces, or other samples, on specialist medium.
- Identification by morphology and biochemical testing.

Treatment

Diarrhoea is often self-limiting, but patients may be treated with erythromycin or fluoroquinolones. An aminoglycoside may be added for patients who have septicaemia.

Prevention and control

Prevention of campylobacteriosis depends on good animal husbandry and abattoir practices, and on good food hygiene in shops, dairies and the home.

Helicobacter pylori

Helicobacter pylori is a motile Gram-negative, spiral bacillus. H. cinaedi and H. fennelliae have been isolated from patients with HIV infection complicated by proctocolitis and bacteraemia.

Pathogenesis

Helicobacter pylori expresses urease, which raises the pH in the surrounding locality, therefore protecting the bacterium from the effects of gastric acid. The CagA pathogenicity island encodes a type IV secretion system that injects the CagA protein into the host cytoplasm, where subversion of a variety of cellular functions occurs that results in IL-8 secretion and inflammatory cell recruitment. VacA, a secreted protein that damages cells, is associated with severe disease.

Clinical features

- Often asymptomatic.
- Chronic infection often takes the form of a low-grade gastritis.
- Strongly associated with both gastric and duodenal ulceration.
- Associated with an increased risk of gastric cancer.

Diagnosis

- Culture or nucleic acid amplification test (NAAT) of gastric and duodenal biopsy.
- Urease breath test.
- Stool antigen test.
- Serology, but this cannot distinguish between recent or old infection.

Treatment

- Proton pump inhibitor, clarithromycin and metronidazole or amoxicillin.
- Specialist advice is required if resistance is suspected.
- Re-infection with H. pylori in adulthood is unusual.

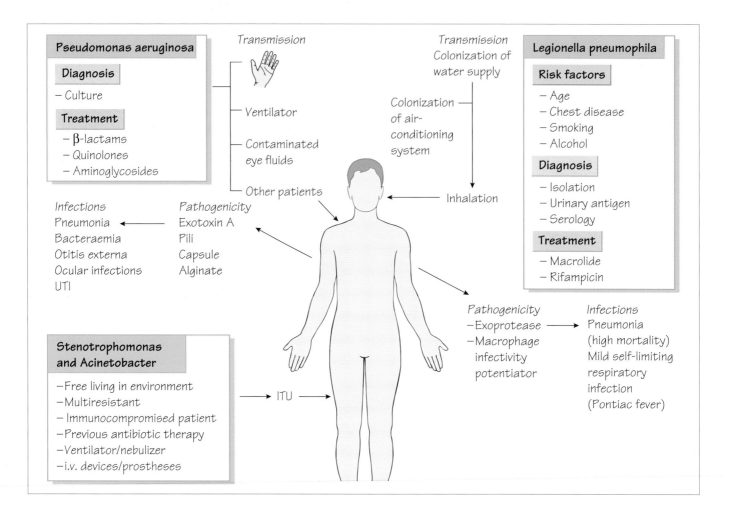

Pseudomonas spp.

Most *Pseudomonas* spp. are environmental organisms that can cause opportunistic infections in a healthcare environment.

Pseudomonas aeruginosa

This organism is widespread in the environment, but rare in the flora of healthy individuals. Its carriage increases with hospitalization. Moist places such as sink-traps, drains and flower vases can harbour *Pseudomonas*.

Pathogenesis

• Produces cytotoxins and proteases (e.g. exotoxins A and S, haemolysins and elastase).
• Isolates from patients with cystic fibrosis produce a polysaccharide alginate that protects from opsonization, phagocytosis and antibiotics in microcolonies.
• Alginate, pili and outer membrane protein mediate adherence.
• Alginate production is associated with hypersusceptibility to antibiotics, lipopolysaccharide deficiency, non-motility and reduced exotoxin production.

Clinical syndromes

• Chronic pulmonary infection in cystic fibrosis.
• Septicaemia, which has a high mortality and is a particular threat to neutropenic patients.
• Rapidly progressive corneal infection and otitis externa.
• Colonization of burns, followed by septicaemia.
• Ecthyma gangrenosum, a destructive skin complication of bacteraemia.
• Osteomyelitis, septic arthritis and meningitis.

Laboratory diagnosis

• Culture on selective media containing cetrimide, irgasin and naladixic acid.
• Identification by biochemical testing.
• Typing by pulse-field gel electrophoresis or multilocus sequence typing (MLST).

Treatment

Organisms are often resistant; therefore, treatment is guided by susceptibilities.

Medical Microbiology and Infection at a Glance, Fourth Edition. Stephen H. Gillespie, Kathleen B. Bamford. © 2012 John Wiley & Sons, Ltd.

Prevention and control

Despite active research there is no effective vaccine available; transmission of multiresistant strains should be controlled by the methods described in Chapter 10.

Burkholderia spp.

Burkholderia cepacia

- This organism can cause chronic pulmonary infection among patients with cystic fibrosis.
- It can spread from patient to patient in cystic fibrosis clinics.
- It is naturally resistant to many antibiotics.
- Treatment is based on susceptibility tests and may include expanded-spectrum cephalosporins, carbapenems or ureidopenicillins.

Burkholderia pseudomallei

This organism is found in soil and water in the tropics. It causes melioidosis that may present as a tuberculosis-like disease, as acute septicaemia or as multiple abscesses. Septicaemia is associated with a high mortality. The diagnosis is made by cultivating the organism from blood or tissues. Treatment is with ceftazidime or imipenem. *B. mallei* causes a similar infection in horses, known as glanders, which can spread to humans.

Stenotrophomonas maltophilia and Acinetobacter spp.

These organisms are found in moist environments, are naturally resistant to many antibiotics and can colonize patients who are immunocompromised or in intensive care units Infection is transmitted by staff or by contaminated shared equipment, such as nebulizers, and is more likely to occur in patients who are receiving antibiotics, have multiple cannulae or are intubated. Both organisms have been implicated in outbreaks of multidrug-resistant infection and systemic invasion leads to pneumonia, septicaemia, meningitis or urinary tract infection. Treatment, when indicated, is based on the results of susceptibility tests.

Legionella spp.

- There are more than 39 species of *Legionella*, but *L. pneumophila* is most frequently implicated in human disease.
- They are found in rivers, lakes, warm springs, domestic water-supplies, fountains, air-conditioning systems, swimming pools and jacuzzis.
- Multiplication occurs at temperatures between 20 and 40 °C inside *Acanthamoeba*.
- Transmission is via aerosols generated from, for example, showers and air-conditioning systems.
- Infection is associated with previous lung disease, smoking and high alcohol intake, but previously healthy patients can be infected.
- Immunocompromised patients in hospital are vulnerable to infection if the hospital air-conditioning system is not adequately maintained.

Pathogenesis

- A major outer membrane protein that inhibits acidification of the phagolysosome.
- Macrophage infectivity is required for optimal internalization.
- *L. pneumophila* expresses a potent exoprotease.

Clinical features

- A mild, influenza-like illness (Pontiac fever).
- Severe pneumonia (Legionnaires' disease), which can lead to respiratory failure and high mortality.
- Patients may complain of nausea or vomiting and malaise before lung symptoms become prominent.
- Cough, which is usually unproductive, and dyspnoea, which is progressive.
- Confusion is common.
- Inappropriate naturetic hormone production may be associated with low serum sodium.

Laboratory diagnosis

- Culture of sputum or, preferably, bronchoalveolar lavage fluid.
- Rapid diagnosis by antigen detection in urine.
- Direct immunofluorescence or nucleic acid amplification test (NAAT) of respiratory specimens.
- Serum antibodies can provide a retrospective diagnosis for epidemiological purposes.

Treatment and prevention

Effective regimens usually consist of a macrolide antibiotic together with rifampicin.

Legionellosis is prevented by adequate maintenance of air-conditioning systems and by ensuring that hot-water supplies are above 45 °C to prevent multiplication.

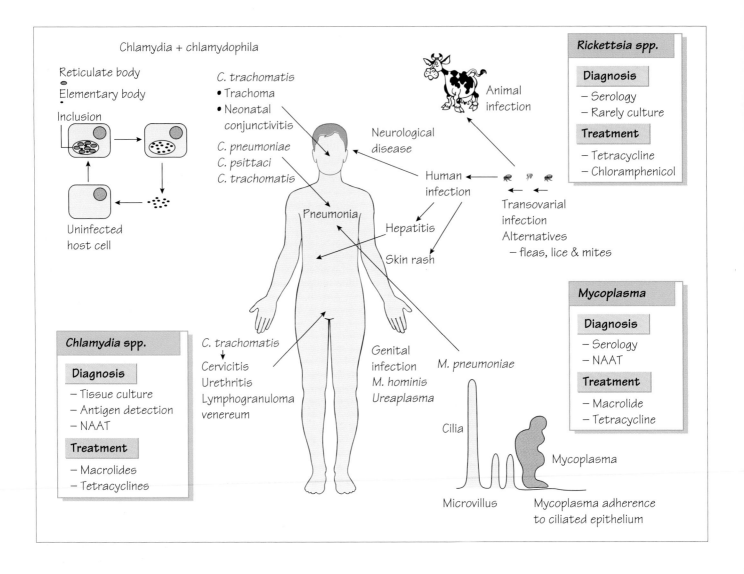

Chlamydia

There are three species: *Chlamydia trachomatis*, which infects the eye and the genital tract; and two respiratory pathogens, *C. pneumonia* and a related species *Chlamydophila psittaci*. They are obligate, intracellular bacteria that exist in two forms: the reticulate body (a non-infective, intracellular, vegetative form) and the elementary body (an extracellular form that permits the organism to survive and be transmitted), which is derived from the reticulate body by binary fission.

Pathogenicity

The major outer membrane protein may participate in attachment to mucosal cells. A 60-kDa cysteine-rich protein may also be associated with virulence.

Chlamydia pneumoniae

Chlamydia pneumoniae is transmitted from person to person by the respiratory route. It produces pneumonia or bronchitis, which is usually clinically mild, but may be associated with pharyngitis, sinusitis and laryngitis.

Chlamydia trachomatis

There are multiple serotypes of *C. trachomatis*: A–C are associated with trachoma (see Chapter 56) and neonatal conjunctivitis (see Chapter 45); D–K are associated with acute urethritis and pelvic inflammatory disease (see Chapter 51); serotypes L1–L3 are associated with lymphogranuloma venereum (see Chapter 51).

Medical Microbiology and Infection at a Glance, Fourth Edition. Stephen H. Gillespie, Kathleen B. Bamford. © 2012 John Wiley & Sons, Ltd.

Chlamydophila psittaci

Chlamydophila psittaci, a pathogen of birds and mammals, causes psittacosis in humans. After an incubation period of 10–14 days, fever, a dry unproductive cough, dyspnoea, headache and myalgia develop. Clinical examination reveals few signs of consolidation, but chest X-ray may show patchy consolidation.

Laboratory diagnosis

• For *C. trachomatis*, usually by nucleic acid amplification test (NAAT) although enzyme immunoassay (EIA) and culture are available.
• For *C. pneumoniae*, usually by NAAT or serology.
• Psittacosis is usually diagnosed serologically.

Mycoplasma and Ureaplasma

Mycoplasma and *Ureaplasma* are small bacteria that lack a cell wall.
• *Mycoplasma pneumoniae*, a human respiratory pathogen, is the second most common cause of respiratory infection after *Streptococcus pneumoniae*.
• *Mycoplasma hominis* and *Ureaplasma urealyticum* have uncertain roles in relation to genital infection.
• *Mycoplasma genitalium* may be associated with urethritis.

Pathogenicity

Mycoplasma pneumoniae adheres to host cells by the P1 protein, localizing to the base of the cilia where it induces ciliostasis. Secreted hydrogen peroxide damages host membranes and interferes with superoxide dismutase and catalase. Opsonized *M. pneumoniae* is readily killed by macrophages and by the activity of the complement system.

Clinical features

Patients present with fever, myalgia, pleuritic chest pain and a non-productive cough; headache is a prominent symptom. Antibodies that agglutinate the host's red blood cells at low temperature (cold agglutinins) cause peripheral and central cyanosis. Infection is associated with reactive (postinfective) arthritis, and neuritis.

Laboratory diagnosis

NAAT is the method of choice for all *Mycoplasma* spp., although serology is an alternative.

Treatment

• They are sensitive to erythromycin, tetracycline, aminoglycosides, rifampicin, chloramphenicol and quinolones.
• They are resistant to β-lactams.

Rickettsia

These organisms are obligate intracellular bacteria with biochemical similarities to Gram-negative bacteria. Clinically, they are divided into three groups:
1 spotted fever;
2 scrub typhus;
3 typhus.

Spotted fever is generally transmitted by ticks. Scrub typhus is caused by a single species, *Rickettsia tsutsugamushi*. The typhus group includes *R. prowazekii* and *R. typhi*, which cause epidemic and murine typhus respectively. Typical features include:
• a 14-day incubation period;
• initial non-specific symptoms, which are followed by possible development of fever, arthralgia and malaise, then a rash, conjunctivitis and pharyngitis;
• confusion, which occurs only in a proportion of Rocky Mountain Spotted Fever cases;
• relapse of *R. prowazekii* infection months or years later, which is known as Brill–Zinsser disease and is usually milder than the primary infection;

Diagnosis is usually by NAAT or IgM-specific EIA. Tetracyclines and chloramphenicol are the treatments of choice, but must be initiated early to influence the outcome.

Coxiella burnetii

Coxiella burnetii is a pathogen of cattle, sheep and goats, which localizes in the placenta. It survives desiccation in the environment and is transmitted by contact with infected animals or their products via the aerosol route. Typical features include:
• an acute febrile illness with fever, myalgia and cough;
• a chronic infection, which occurs in approximately 5% of cases;
• Q fever, which may present as atypical pneumonia, pyrexia of uncertain origin and hepatitis.
• various complications including relapses, which can take the form of culture-negative endocarditis or granulomatous hepatitis.

Diagnosis is made by NAAT or EIA and treatment is usually with doxycycline.

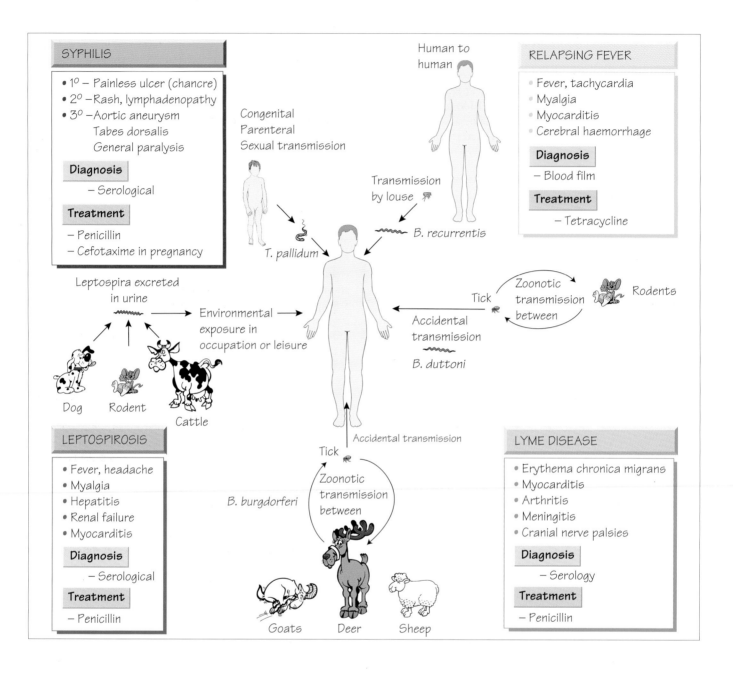

SYPHILIS
- 1° – Painless ulcer (chancre)
- 2° – Rash, lymphadenopathy
- 3° – Aortic aneurysm
  Tabes dorsalis
  General paralysis

Diagnosis
  – Serological

Treatment
– Penicillin
– Cefotaxime in pregnancy

Human to human

Congenital
Parenteral
Sexual transmission

T. pallidum

Transmission by louse

B. recurrentis

RELAPSING FEVER
- Fever, tachycardia
- Myalgia
- Myocarditis
- Cerebral haemorrhage

Diagnosis
– Blood film

Treatment
  – Tetracycline

Leptospira excreted in urine

Environmental exposure in occupation or leisure

Dog   Rodent
Cattle

Tick   Zoonotic transmission between   Rodents

Accidental transmission
B. duttoni

LEPTOSPIROSIS
- Fever, headache
- Myalgia
- Hepatitis
- Renal failure
- Myocarditis

Diagnosis
  – Serological

Treatment
– Penicillin

B. burgdorferi

Accidental transmission

Tick
Zoonotic transmission between

Goats   Deer   Sheep

LYME DISEASE
- Erythema chronica migrans
- Myocarditis
- Arthritis
- Meningitis
- Cranial nerve palsies

Diagnosis
  – Serology

Treatment
– Penicillin

Leptospira

There are two main species: *Leptospira interrogans*, which contains all of the pathogenic strains, and *L. biflexa*, a non-pathogen. *L. interrogans* has more than 200 serovariants that may be written as if they are separate species.

Epidemiology

Leptospira interrogans have different preferred mammalian hosts, for example the rat is the reservoir of *L. interrogans* Icterohaemor-rhagiae. Leptospires colonize the renal tubules of their natural hosts and are excreted in urine. Humans may be infected by contact with animal urine or contaminated water or soil. Water-sports enthusiasts, sewer workers and agricultural workers are at increased risk of infection.

Pathogenesis and clinical features

The central nervous system, liver and kidneys are most affected by the human disease. The severity varies between serovars.

Medical Microbiology and Infection at a Glance, Fourth Edition. Stephen H. Gillespie, Kathleen B. Bamford. © 2012 John Wiley & Sons, Ltd.

There are two phases:
• bacteraemia, with fever, headache, myalgia, conjunctivitis and abdominal pain;
• fever, uveitis and aseptic meningitis, which predominate after the organisms disappear from the blood.

In addition, jaundice, haemorrhage, renal failure and myocarditis occur in severe cases. A poor outcome is associated with hypotension, renal failure and clinical evidence of pulmonary involvement. Diagnosis is made by enzyme immunoassay (EIA), microagglutination or nucleic acid amplification test (NAAT).

Penicillin or doxycycline must be commenced early in the disease. Doxycycline is an effective prophylactic agent if exposure to infection is likely to have occurred.

Borrelia

Borrelia are transmitted to humans via arthropods (lice or ticks) throughout the world, with a well-defined geographical territory and host specificity. For example, humans are the only host of louse-borne relapsing fever (*B. recurrentis*). Epidemics arise during war or mass migration when humans invade the *Borrelia*–tick–rodent habitat.

Relapsing fever

• *Borrelia* cause bacteraemia and fever. Antibodies clear the organism from the blood, but antigenic variation allows relapse. The disease resolves when the repertoire of antigenic variation is exhausted.
• Presentation is with headache, myalgia, tachycardia, rigors, hepatosplenomegaly and a petechial rash.
• Episodes last for 3–6 days; relapses occur a week apart.
• Louse-borne relapsing fever has a high mortality (up to 40%); mortality in tick-borne disease rarely exceeds 5%.
• Myocarditis, cerebral haemorrhage and/or hepatic failure are the usual causes of death.
• Postexposure doxycycline is effective in preventing disease.
• Doxycycline is treatment of choice.

Lyme disease

Borrelia burgdorferi, *B. afzelii* and *B. garinii* are transmitted by *Ixodes* ticks and cause Lyme disease, which is endemic in the eastern USA and Europe. Humans are accidental hosts. The early symptoms are caused by the acute infective process; later manifestations are thought to be related to the host immune response. Typical features include:
• an initial expanding red macule or papule (erythema chronicum migrans);
• later headache, conjunctivitis, fever and regional lymphadenopathy;
• complications including new skin lesions, myocarditis, arthritis, aseptic meningitis, cranial nerve palsies and radiculitis;
• acrodermatitis chronica atrophicans, a red skin lesion, which may also occur.

Diagnosis is made by EIA followed by Western blot. Doxycycline or amoxicillin is used for treatment of early Lyme disease; ceftriaxone for late or recurrent disease.

Vincent's angina

This is a painful, oral, ulcerative, destructive infection with *Borrelia vincentii* and fusobacteria or other anaerobes. Clinical diagnosis is confirmed by Gram stain. Treatment is with penicillin and metronidazole.

Treponema pallidum

Treponema pallidum causes syphilis, which may be transmitted sexually or congenitally. The incidence is now increasing worldwide, having been falling for many years. The related organisms *T. pertenue* and *T. carateum* cause yaws and pinta respectively. They are spread by contact, usually in childhood. Once common in the tropics, they are now rare as the result of an eradication campaign.

Clinical features

Treponema pallidum penetrate intact skin or mucosa disseminating throughout the body to cause disease in four stages:
1 Primary chancre (painless ulcer with a rubbery edge and regional lymphadenopathy).
2 Secondary (an acute febrile illness with a generalized non-itchy scaling rash that typically involves the palms, associated with lymphadenopathy).
3 Latent phase, which may last for many years.
4 Tertiary (systemic lesions become symptomatic, e.g. aortitis, posterior cord degeneration and dementia).

The characteristic syphilitic lesions (gummas), which consist of necrosis and obliterative endarteritis with fibroblastic proliferation and lymphocyte infiltration, are found throughout the body.

Diagnosis

• Primary: dark-ground microscopy or NAAT.
• Later stages: EIA for specific IgG and IgM as well as cardiolipin agglutination (measures disease activity) and tests based on cultivated treponemes such as the treponemal haemagglutination test (TPHA).
• CSF testing should be performed to detect early central nervous system involvement.
• Neonatal: detection of IgM by Western blotting.

Treatment

• Penicillin (or either azithromycin or tetracyclines if the patient is allergic).
• An acute febrile response (the Jarisch–Herxheimer reaction) may develop in some patients after the first dose of antibiotics.
• Careful serological follow-up is essential to confirm cure and/or detect early central nervous system involvement.

29 Virus structure, classification and antiviral therapy

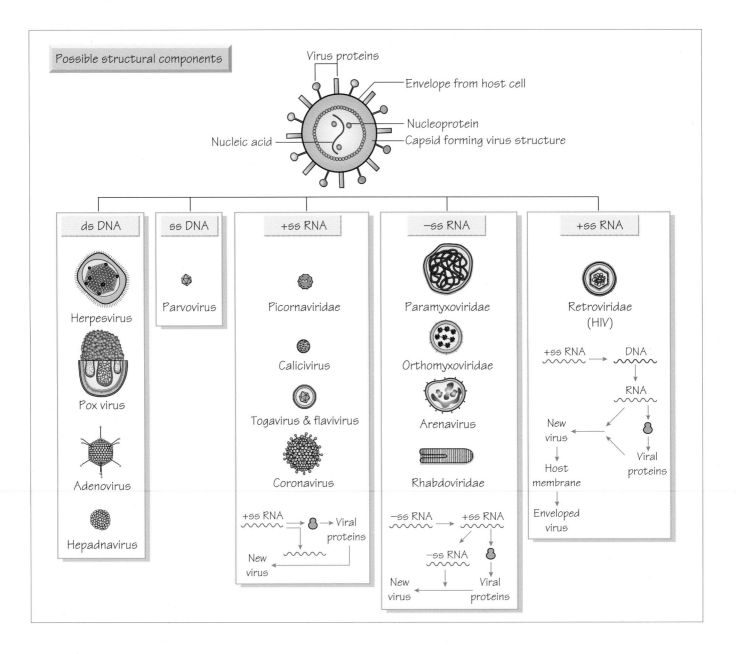

Viral classification

Viral classification is based on the nucleotides in the virus, its mode of replication, the structure and symmetry of the structural proteins (capsids) and the presence or absence of an envelope.

Genetic material and replication

DNA viruses

• Double-stranded DNA viruses include poxviruses, herpesviruses, adenoviruses, papovaviruses and polyomaviruses.

• Single-stranded DNA viruses include parvoviruses.

DNA viruses usually replicate in the nucleus of host cells by producing a polymerase that reproduces viral DNA. Viral DNA is not usually incorporated into host chromosomal DNA.

RNA viruses

RNA viruses possess a single strand of RNA and adopt different reproductive strategies:

• RNA sense (positive) may serve directly as mRNA and be translated into structural protein and an RNA-dependent RNA polymerase.

Medical Microbiology and Infection at a Glance, Fourth Edition. Stephen H. Gillespie, Kathleen B. Bamford. © 2012 John Wiley & Sons, Ltd.

• RNA antisense (negative) contains an RNA-dependent RNA polymerase that transcribes the viral genome into mRNA. Alternatively, the transcribed RNA can act as a template for further viral (antisense) RNA.

• Retroviruses have single-stranded sense RNA that cannot act as mRNA. This is transcribed into DNA by reverse transcriptase and incorporated into host DNA. The subsequent transcription to make mRNA and viral genomic RNA is under the control of host transcriptase enzymes.

Capsid symmetry

Viral nucleic acid is covered by a protein coat of repeating units (capsids), with either icosahedral (spherical) or helical (arranged around a rotational axis) symmetry.

Repeating units reduce the number of genes devoted to production of the viral coat and simplify the process of viral assembly.

Envelope

A lipid envelope derived from host cell or nuclear membrane surrounds some viruses. The host membrane may incorporate viral-encoded antigens that may act as receptors for other host cells. Enveloped viruses are sensitive to substances that dissolve the lipid membrane (e.g. ether).

Antiviral therapy

The intracellular location of viruses and their use of host cell systems pose a challenge to the development of antiviral therapy. Drugs may work at different stages of viral replication.

Uncoating

Amantadine/rimantidine prevents uncoating and release of influenza RNA but resistance arises readily. Pleconaril inhibits uncoating of picornaviruses and is active against enteroviruses and rhinoviruses; it is absorbed orally and clinical trials suggest it shortens clinical symptoms.

Nucleoside analogues
Chain termination
Aciclovir is selectively converted into acyclo-guanosine monophosphate (acyclo-GMP) by viral enzymes, then into a potent inhibitor of viral DNA polymerase by host enzymes. The acyclo-GMP causes viral DNA **chain termination**. Resistance occurs through the development of deficient thymidine kinase production or alteration in the viral polymerase gene. The drug can be taken orally and crosses the blood–brain barrier. Other agents (e.g. ganciclovir) work in a similar way.

Reverse transcriptase inhibition
Lamivudine inhibits the reverse transcriptase of hepatitis B and HIV (see below). Nucleoside and nucleotide inhibitors are being developed as alternative treatments for hepatitis B; these include adefovir, entecavir, tenofovir, telbivudine and clevudine.

Ribavirin is a guanosine analogue that inhibits several steps in viral replication including capping and elongation of viral mRNA. It is active against respiratory syncytial virus, influenza A and B, parainfluenza virus, Lassa fever, hantavirus and other arenaviruses.

Nucleoside reverse transcriptase inhibitors
Nucleoside reverse transcriptase inhibitors (NRTIs) inhibit reverse transcriptase by being incorporated as faulty nucleotides. Examples include the longest established antiretroviral drug zidovudine (AZT), plus lamivudine (3TC), stavudine (d4T), tenofovir, didanosine (ddI) zalcitabine (ddC) and abacavir (see Chapter 46).

Non-nucleoside reverse transcriptase inhibitors
Non-nucleoside reverse transcriptase inhibitors (NNRTIs) inhibit reverse transcriptase directly; examples include nevirapine, efavirenz, delavirdine and etravirine. They have been shown to be effective agents in combination regimens. As resistance occurs after a single mutation, they are used in maximally suppressive regimens only.

Protease inhibitors
Protease inhibitors target the HIV-encoded protease. They are highly effective antiretroviral compounds that cause significant falls in viral load. They include atazanavir, indinavir, lopinavir, ritonavir and saquinavir. Ruprintrivir acts in the same way against human rhinovirus 3C protease. It is administered by nasal spray and appears to have useful activity in rhinovirus infection.

Fusion inhibitors
Enfuvirtide inhibits binding with gp134; maraviroc inhibits binding to CCR5 preventing fusion. Both agents are used for salvage therapy in AIDS (see Chapter 46).

Release inhibitors
Neuraminidase inhibitors including zanamivir and oseltamivir inhibit the final stage in the release of virus from the host cell.

Integrase inhibitors
These agents are being developed to block the insertion of the HIV viral genome into the DNA of the host cell.

Other agents
Infections with hepatitis B and hepatitis C can also be treated with α-interferon, a host cytokine.

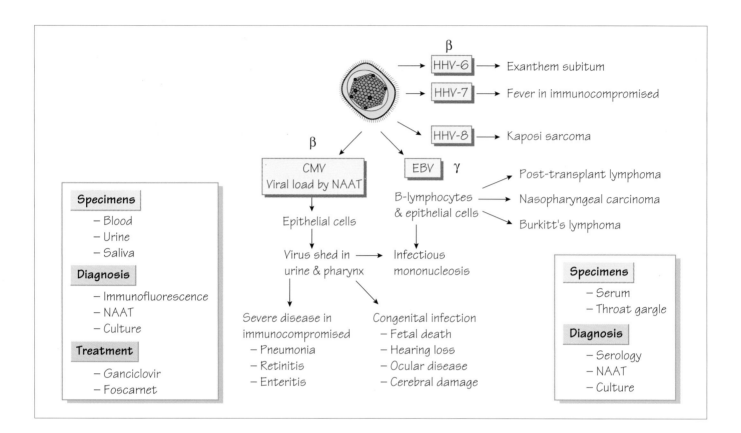

Herpesviruses are enveloped, double-stranded DNA viruses (120–240 kb) encoding for more than 35 proteins. After an acute infection, lifelong latency follows with the potential for relapse to occur later in life, especially if the individual becomes immunocompromised.

Classification

Herpesviruses are divided into three groups:
• α-herpesviruses are fast-growing cytolytic viruses that establish latent infections in neurones (e.g. herpes simplex and varicella zoster);
• β-herpesviruses are slow-growing viruses that become latent in secretory glands and kidneys (e.g. cytomegalovirus [CMV], HHV6 and 7);
• γ-herpesviruses are latent in lymphoid tissues (e.g. Epstein–Barr virus [EBV], HHV-8).

Cytomegalovirus

Epidemiology and pathogenesis
• Transmitted vertically or by close contact.
• Infection occurs later in life with increasing wealth.
• Approximately 50% of adults in the UK have been infected.
• Infection may be transmitted to the fetus before or after birth.
• Infection may also be acquired from blood transfusion or organ transplantation.

Clinical features
• Neonatal infection can be severe (see Chapter 45), or may be initially asymptomatic, later leading to the development of deafness and/or to developmental milestone delay.
• Postnatal infection is usually mild.
• Immunocompromised patients, especially those with HIV infection or who have undergone organ transplantation, may develop severe pneumonitis, retinitis or gut infection through reactivation of latent infection or infection from the donor organ.

Diagnosis
• Diagnosis is usually by nucleic acid amplification test (NAAT) on blood, urine or respiratory samples.
• Monitoring of viral load is important to identify patients with severe disease who require treatment.
• The virus is readily cultured.

Treatment and prevention
• Severe infections that threaten life or sight should be treated with ganciclovir, together with immunoglobulin in the case of pneumonitis.
• Valganciclovir, the ester of ganciclovir, is an oral preparation used for initial treatment and maintenance.
• Alternatives, all of which are more toxic, include foscarnet and cidofovir, a DNA polymerase chain inhibitor.

Medical Microbiology and Infection at a Glance, Fourth Edition. Stephen H. Gillespie, Kathleen B. Bamford. © 2012 John Wiley & Sons, Ltd.

- Appropriate screening of donor organs and blood products can reduce the risk of transmission.

Epstein–Barr virus
Epidemiology and pathogenesis
As with CMV, infection is generally found in the very young in developing countries and in adults in industrialized countries. Gaining entry via the pharynx, the virus infects B cells and disseminates widely. EBV is capable of immortalization of B cells causing neoplasia: Burkitt's lymphoma (found in sub-Saharan Africa in association with malaria); nasopharyngeal carcinoma (in China); and lymphoma (in immunocompromised patients including transplant recipients).

Clinical features
- Infection is characterized by fever, malaise, fatigue, sore throat, lymphadenopathy and, occasionally, by hepatitis.
- Symptoms usually last about 2 weeks.
- Persistent symptoms may develop in a few patients.
- EBV infection is associated with tumours (see above).

Diagnosis
- Rapid slide agglutination technique.
- Definitive diagnosis is by detection of specific IgM to EBV viral capsid antigen.
- NAAT-based diagnosis can now also be used.
- The pattern of immune response to Epstein–Barr nuclear antigen complex (EBNA), latent membrane protein, terminal protein, the membrane antigen complex and the early antigen (EA) complex allow the stage of infection to be determined.

Human herpesviruses 6 and 7
- The sole member of the *Roseolovirus* genus, herpesvirus 6 (HHV-6) has two subtypes, A and B, which infect human T cells.
- Transmission is probably through infected saliva; almost all individuals are infected by the end of their second year.
- The infection, known as 'exanthem subitum', is characterized by a 3 to 5-day febrile illness that settles as the rash appears.
- Asymptomatic infection is common.
- It may be associated with febrile convulsions and encephalitis, although the latter is rare.
- Hepatitis is another rare complication.
- An IgG enzyme immunoassay (EIA) is available and a quantitative NAAT may be helpful in the diagnosis.
- Infection with HHV-7 is almost universal by the age of 5, but there is no clear association with disease.
- Diagnosis is with paired sera to detect antibody levels.

Human Kaposi sarcomavirus or human herpesvirus 8
The human Kaposi sarcomavirus (HHV-8) is a γ-herpesvirus. Transmission can be vertical from mother to child, and in the young is by mucosal (non-sexual) contact. Initial infection is characterized by infectious mononucleosis-like syndrome. Later, immunocompromised patients, especially those with AIDS, may develop Kaposi sarcoma. Diagnosis is principally by NAAT in suspect tissues. Serological tests using EIA and indirect fluorescence are available.

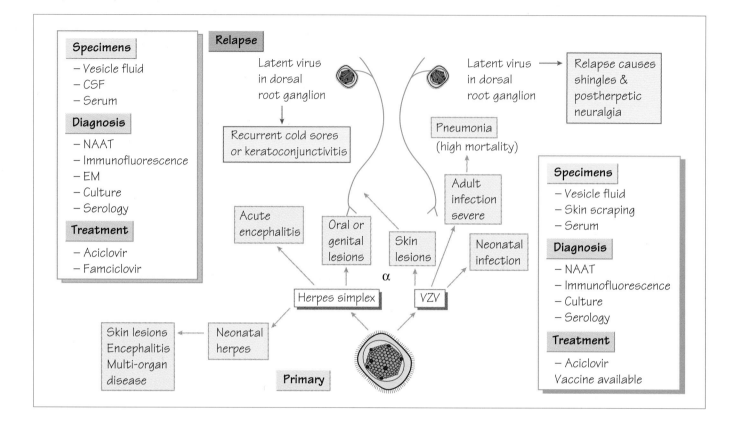

Herpes simplex

Pathogenesis and epidemiology

- Transmitted by direct contact.
- Invades skin locally producing skin vesicles by its cytolytic activity.
- Remains latent in the sensory ganglia.
- Reactivation is triggered by physical factors (e.g. infection, sunlight), or psychological stress.
- Cell-mediated immunity controls infection, therefore immunocompromised patients are at risk of reactivation and severe infection.

Clinical features

- Herpes simplex virus 1 (HSV-1) is often asymptomatic, but young children commonly develop fever, vesicular gingivostomatitis and lymphadenopathy.
- Adults with infection may exhibit pharyngitis and tonsillitis.
- Primary eye infection produces severe keratoconjunctivitis; recurrent infection may result in corneal scarring.
- Primary skin infection (herpetic whitlow) usually occurs in traumatized skin (e.g. on fingers).
- Severe encephalitis may occur (see Chapter 49).

- Mother-to-child transmission perinatally may result in a generalized neonatal infection including encephalitis.
- HSV-2 infection causes painful genital ulceration that lasts up to 3 weeks and is associated with recurrence.
- Genital herpes is an important cofactor in the transmission of HIV.
- Meningitis is an uncommon complication of primary type 2 infection.

Diagnosis

A nucleic acid amplification test (NAAT) of vesicle fluid, genital or mouth swabs is the standard diagnostic method, although the virus grows readily and can be visualized by electron microscopy (EM). The ratio between serum and CSF antibody may indicate local production and can help in the diagnosis of HSV encephalitis. MRI or CT scans of the brain may detect temporal lobe lesions that are typical of herpes encephalitis.

Treatment

Topical, oral and intravenous preparations of aciclovir and other agents with better oral absorption, including valaciclovir and famciclovir, are available. Encephalitis is treated with intravenous aciclovir.

Medical Microbiology and Infection at a Glance, Fourth Edition. Stephen H. Gillespie, Kathleen B. Bamford. © 2012 John Wiley & Sons, Ltd.

Varicella zoster virus

Varicella zoster virus (VZV), which has only one serological type, causes the acute primary infection known as **chickenpox** and its recurrence, which is called **shingles**.

Pathogenesis and epidemiology
- VZV is found in the vesicle and transmission is by contact and airborne spread from patients with vesicles.
- The attack rate in non-immune individuals is very high (>90%).
- The incubation period is 14–21 days.
- Infection is commonest in children aged 4–10 years.
- Recovery provides lifelong immunity.
- The virus remains latent in the posterior root ganglion and in 20% of patients will reactivate with lesions in the related dermatome, causing shingles.
- Shingles lesions contain VZV and are infectious to non-immune individuals who are at risk of developing chickenpox.
- It is impossible to contract shingles directly from chickenpox or other cases of shingles.

Clinical features
- The discomfort of chickenpox comes from the rash.
- Systemic symptoms are mild.
- Lesions, which appear in crops usually 2 or 3 days apart, affect all parts of the body, including the oropharynx and genitourinary tract, and progress through macules and papules to vesicular eruptions which, following rupture, develop a crust and spontaneously heal.
- The rash lasts for 7–10 days, but complete resolution may take as long again.
- Haemorrhagic skin lesions that can be life-threatening may occur.
- Secondary infection with *Staphylococcus aureus* or *Streptococcus pyogenes* may also require treatment.

- VZV pneumonia is more common in adults, especially in immunocompromised individuals, and has a high mortality; survivors may recover completely or may have respiratory impairment.
- Postinfectious encephalitis, which is usually minor, can occur, but there is also a rare fatal form.
- Maternal transmission through contact with vaginal lesions during birth can result in severe neonatal infection.
- Shingles is a painful condition that usually affects older people or immunocompromised individuals.
- Ocular damage may follow the involvement of the ophthalmic division of the trigeminal nerve.
- Up to 10% of shingles episodes will be followed by postherpetic neuralgia, a very painful condition that may last for many years and can be associated with suicide.

Diagnosis
- Both chickenpox and shingles are usually diagnosed clinically.
- Laboratory diagnosis is by NAAT.
- Staining of fluid from a vesicle may show characteristic giant cells.
- VZV may be visualized by EM or cultured.
- Serology is important to determine the immune status of patients and staff in outbreaks.

Treatment and prevention
- Aciclovir or valaciclovir may be used for both adult chickenpox and shingles.
- Postherpetic neuralgia may be reduced by early treatment.
- Pain may be severe and require referral to a pain clinic.
- A live attenuated-virus vaccine is available and recommended for non-immune healthcare workers.
- Zoster immune globulin (ZIG) is given to those in close contact with infection who are at risk of serious disease (e.g. neonates, pregnant women and immunocompromised individuals).

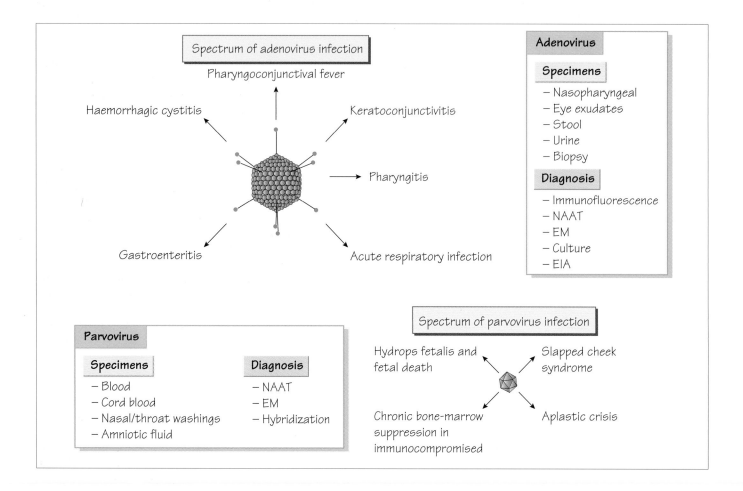

Adenovirus

Adenoviruses are unenveloped, icosahedral, double-stranded DNA viruses that possess species-specific, group-specific and type-specific antigens. There are more than 50 serotypes of human adenoviruses, which are divided into six groups (A–F) on the basis of their genomic homology.

Epidemiology and clinical features

- Transmitted by direct contact and faecal–oral route.
- Pharyngoconjunctival fever is caused by serotypes 3 and 7.
- Acute febrile pharyngitis is caused by serotypes 1–7.
- Serotypes 40 and 41 cause enteric infection.
- Serotypes 8, 19 and 37 cause conjunctivitis.
- Serotypes 4, 17 and 14 cause respiratory infection.
- Haemorrhagic cystitis is caused by serotypes 11 and 21.
- Immunocompromised patients may suffer severe pneumonia (serotypes 1–7), urethritis (serotype 37) and hepatitis in liver allografts.
- The clinical spectrum may vary depending on the site of infection.

Genetically modified adenoviruses and adeno-associated viruses are increasingly being explored as vectors for gene therapy.

Diagnosis

Diagnosis is usually made by nucleic acid amplification test (NAAT) but culture, serology and electron microscopy (EM) diagnosis are available.

Prevention and control

Outbreaks must be managed according to infection control practices (both respiratory and contact). Outbreaks of ocular infection at swimming pools are prevented by adequate chlorination. Transmission between patients undergoing ophthalmic examination can be prevented by single-use equipment, adequate decontamination of equipment and appropriate hygiene by healthcare staff.

Parvovirus

Parvoviruses are small, unenveloped, icosahedral, single-stranded DNA viruses with one serotype, B19, known to cause human disease and given the genus name *Erythrovirus*.

Medical Microbiology and Infection at a Glance, Fourth Edition. Stephen H. Gillespie, Kathleen B. Bamford. © 2012 John Wiley & Sons, Ltd.

Epidemiology

Infection is found worldwide and throughout the year. Transmission is by the respiratory route. It may cause outbreaks of erythema infectiosum in schools. Seroprevalence increases with age with more than 60% of adults possessing antibody.

Pathogenesis and clinical features

• Parvovirus B19 invades red blood cells through globoside P replicating in immature erythrocytes.
• It produces erythema infectiosum, a mild febrile disease that typically occurs in young children who may exhibit a 'slapped cheek' appearance.
• A symmetrical, small-joint arthritis may also develop, especially in adults.
• Red cell production is arrested by infection, which may cause severe anaemia in patients with a high red blood cell turnover (e.g. aplastic crises in patients with sickle cell disease).
• The risk of infection in pregnancy is low, but it may lead to hydrops fetalis and fetal death, although there is no evidence that parvovirus causes congenital abnormalities.
• Infection during the first 20 weeks of pregnancy results in 10% fetal loss.

Diagnosis

• Diagnosis is usually made clinically, but NAAT is the test of choice.
• Detection of IgM is also used.
• Blood, nasal or throat washings, cord blood and amniotic fluid can be examined by EM.

Prevention and control

No specific treatment or vaccine is available at present.

Respiratory precautions should prevent transmission in the hospital environment.

Papillomavirus

These are small, enveloped, double-stranded DNA viruses with more than 100 types. Some are responsible for common warts and genital warts. Types 16 and 18 predominate in cervical neoplasia; they are transmitted by close contact, including by the sexual route. Diagnosis of a common wart is clinical; cervical neoplasm is diagnosed by cytology and NAAT. A vaccine against types 6, 11, 16 and 18 is now in use.

Poxvirus

Poxviruses are double-stranded DNA viruses with complex symmetry and a shape that resembles a ball of wool.

Smallpox

Once a major cause of death worldwide this has now been eradicated but there are concerns that smallpox may become a bioterrorism weapon, which have prompted some countries to produce stocks of vaccine.

Monkeypox

A zoonotic infection in rainforest areas of Central and West Africa that is similar to smallpox. The case fatality rate can reach 10% in Africa, but was much lower in the USA where there was an outbreak associated with infected prairie dogs. Diagnosis is by EM or NAAT.

Orf

A zoonotic, pustular dermatitis originating in sheep and goats that is characterized by a single vesicular lesion, which is typically found on the finger and resolves spontaneously after a few weeks. Diagnosis is usually clinical on the basis of appearance and a history of exposure.

Molluscum contagiosum

• A common condition, especially in children, with crops of small, regular, papular, 'pearl-like' skin lesions, usually occurring on the face, arms, buttocks and back.
• It may be transmitted sexually, by direct contact or on fomites.
• Steroid therapy and/or infection with HIV increase the extent of disease.
• The microscopic appearance is of epidermal hypertrophy that extends into the dermis, and cells with inclusion bodies that are seen in the prickle-cell layer.
• Diagnosis is usually clinical and can be confirmed by EM examination of lesion scrapings.
• The rash may last 1 year in immunocompetent individuals and may become a chronic problem for patients with HIV infection. Traditional treatment – by prodding the lesions with a sharp implement – promotes healing.

Tanapox

Tanapox is a febrile illness usually associated with a single nodular skin lesion that may ulcerate and heal spontaneously. Infection is acquired in central and east Africa; the diagnosis is usually suggested by the travel history and can be confirmed by EM or NAAT.

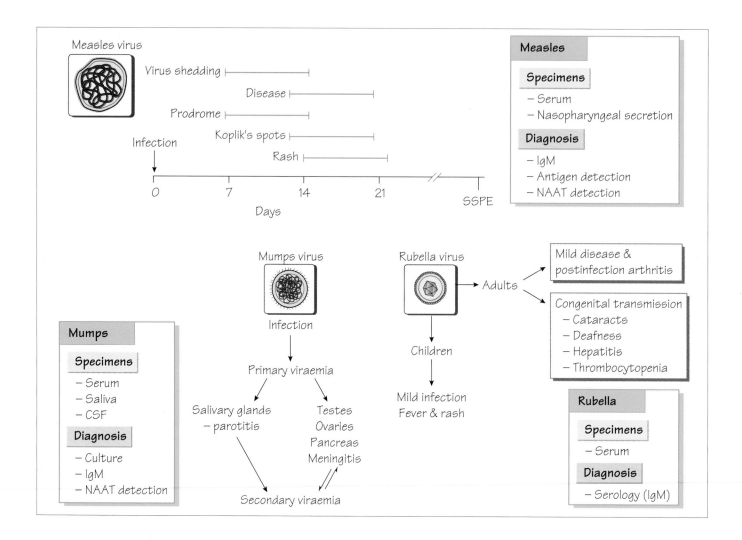

Measles

Measles is due to an enveloped RNA virus, known as a *Morbillivirus*, with a single serotype. The virus encodes six structural proteins that facilitate attachment to the host cell and viral entry, which includes two transmembrane glycoproteins: fusion (F) and haemagglutinin (H). Antibodies to F and H are protective.

Pathogenesis and epidemiology

• Initially the virus infects epithelial cells of the upper respiratory tract.

• It then invades neighbouring lymphoid tissue, which results in primary viraemia and involvement of the reticuloendothelial system.

• This is followed by a secondary viraemia and dissemination throughout the body, which coincides with the onset of clinical symptoms.

• It is transmitted by the airborne route, with a high attack rate.

• The incubation period is 9–12 days – individuals are infectious for 3 days before the rash emerges.

• Natural infection is followed by lifelong immunity.

• Mortality is rare except in patients who have HIV infection, are immunocompromised or malnourished (especially those with vitamin A deficiency); mortality rates are highest in children under 2 years of age.

• Measles is rare in countries with a vaccination programme but 90% coverage is required to ensure the disease does not re-emerge.

Clinical features

• A prodromal 2 to 4-day coryzal illness occurs, during which small white papules (Koplik's spots) are found on the buccal mucosa near the first premolars.

• A morbilliform rash appears, first behind the ears, then spreading centrifugally and becoming brownish.

• Secondary pneumonia, otitis media and croup are common complications.

• Acute postinfectious encephalitis is a rare and serious complication.

Medical Microbiology and Infection at a Glance, Fourth Edition. Stephen H. Gillespie, Kathleen B. Bamford. © 2012 John Wiley & Sons, Ltd.

- Subacute encephalitis, a chronic progressive disease, occurs mainly in children with leukaemia.
- Subacute sclerosing panencephalitis (SSPE) is a rare, progressive, fatal encephalitis that develops more than 6 years after infection.

Diagnosis
- Diagnosis is usually clinical, but may be confirmed by salivary IgM-specific enzyme immunoassay (EIA).
- SSPE is diagnosed by detection of virus-specific antibody that is being synthesized in the CSF (e.g. specific IgM).
- A nucleic acid amplification test (NAAT) and molecular characterization of the virus by sequencing are also available.

Mumps
A member of the *Paramyxovirus* genus, the mumps virus is a pleomorphic, enveloped, antisense RNA virus with one serotype.

Epidemiology
- Mumps usually occurs in childhood but many adults are susceptible as it has a relatively low attack rate.
- The incubation period is 14–24 days.
- Subclinical infection is common, especially in children.
- It is transmitted readily by the aerial route.
- Infection creates lifelong immunity.
- Epidemics can re-emerge if vaccination coverage falls.

Clinical features
- Common features include fever, malaise, myalgia and parotid gland inflammation.
- Meningitis occurs in up to 15% of patients with parotitis.
- Complete recovery is almost invariable, although rare fatal forms and postmeningitis deafness may occur.
- Complications include orchitis (20%), oophoritis (5%) or pancreatitis (5%) usually in older individuals.

Diagnosis
- Diagnosis is usually clinical, but may be confirmed by specific salivary or serum IgM.
- NAAT for diagnosis is also available.

Rubella
Rubella (rubivirus), which is a member of the Togaviridae family, is an icosahedral, pleomorphic, enveloped, positive-strand RNA virus with a single serotype.

Epidemiology
- Rubella is rare in countries with a vaccination programme.
- Transmission is by aerial droplets.
- Patients are infectious from 7 days before the rash appears until 14 days after the rash.
- Natural infection is followed by solid immunity.

Clinical features
Rubella is associated with fever, a fine, red, maculopapular rash and lymphadenopathy. During the prodrome red pinpoint lesions occur on the soft palate. Arthritis (more common in females) and self-limiting encephalitis are complications.

Maternal infection may cause fetal death or severe abnormalities, such as deafness, central nervous system deficit, cataract, neonatal purpura and cardiac defects, in up to 60% of cases; the risk being highest during the first trimester.

Diagnosis
- Diagnosis is by detection of IgM and IgG antibodies in serum or saliva.
- Congenital disease is diagnosed by finding specific IgM persistent antibodies (>6 months) in an infant, or viral detection by culture or NAAT.

Prevention of measles, mumps and rubella
- A live attenuated combined vaccine (the MMR) is given between 13 and 15 months, with a booster dose given at school entry.
- Further booster doses of measles vaccine may be required.
- The rapid antibody response to measles vaccine can be used to protect susceptible individuals exposed to measles.
- Women attending for contraceptive advice should be screened for rubella antibodies and vaccinated if not pregnant.
- MMR should not be given to immunocompromised individuals.

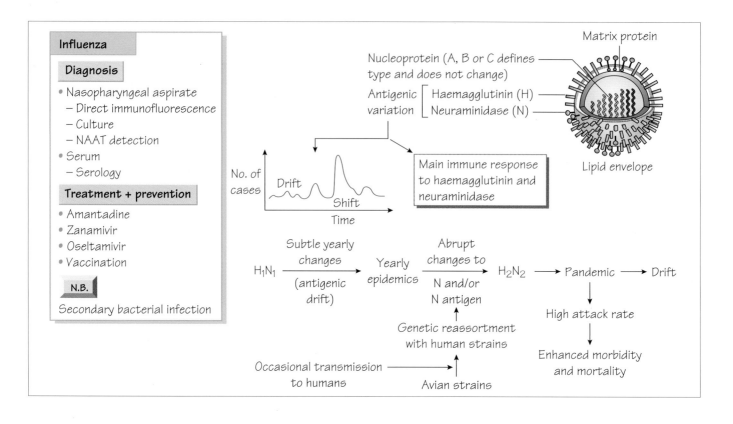

Influenza virus

Virology and epidemiology

Influenza virus is an enveloped orthomyxovirus (100 nm) that contains a negative single-stranded RNA genome divided into eight segments. This structure facilitates genetic re-assortment, which allows the virus to change its surface antigens and the influenza virus will take up genetic material from avian and pig influenza strains. The virus expresses seven proteins, three of which are responsible for RNA transcription. The nucleoprotein has three antigenic types that designate the three main virus groups, influenza A, B and C. Of the three types, influenza A and, more rarely, influenza B undergo genetic shift. The matrix protein forms a shell under the lipid envelope with haemagglutinin and neuraminidase proteins expressed as 10-nm spikes on the envelope, which interact with host cells. Virus immunity is directed against the haemagglutinin (H) and neuraminidase (N) antigens.

Epidemic/pandemic 'flu

Annual epidemics of influenza are possible because the H and N antigens change, known as **antigenic drift**. This means that there are a sufficiently large number of individuals without immunity for the virus to circulate and, in some years, for an epidemic to occur. The virus may also undergo major genetic change, which is often due to gene re-assortment, known as **antigenic shift**. When this happens, as there are very few individuals with immunity, a worldwide pandemic may develop. Pandemics occur every 10–40 years, often originating in the Far East then circulating westwards. Such novel strains can often be traced to infected birds, poultry or pigs. Pandemic influenza A strains have a high attack rate and are associated with increased morbidity and mortality: 20 million people died in the 'Spanish 'flu' epidemic of 1919. The most recent pandemic virus, which arose in Mexico and was designated 'swine 'flu', was an H1N1 virus and had a high attack rate in the young. Viral pneumonia was most common in pregnant women and patients who were immunocompromised, but the global mortality rate was low. The risk of a pandemic is high when there are epizootics of avian 'flu circulating in domestic birds (e.g. H5N1) and genetic re-assortment occurs. Serotypes B and C are exclusively human pathogens that do not cause pandemics.

Avian 'flu

Avian strains are of great concern to poultry farmers, as avian 'flu may cause high mortality in their flocks. Infection can be transmitted to poultry from migratory wild birds. The virus can spread to humans and may be associated with high mortality (e.g. in the case of the H5N1 virus). Person-to-person spread is uncommon.

Clinical features

The incubation period lasts 1–4 days and patients are infectious for approximately 3 days, starting from 1 day before symptoms

Medical Microbiology and Infection at a Glance, Fourth Edition. Stephen H. Gillespie, Kathleen B. Bamford. © 2012 John Wiley & Sons, Ltd.

emerge. Headache, myalgia, fever and cough last for 3–4 days. Complications, which are more common in elderly people and patients with cardiopulmonary disease, include primary viral or secondary bacterial pneumonia.

Diagnosis

Most diagnoses are made clinically. Rapid laboratory diagnosis is by direct immunofluorescence that can detect influenza A/B or C. Nucleic acid amplification tests (NAATs) are more sensitive and can identify the specific serotype, which can indicate whether a patient is infected with the pandemic strain. Public health laboratory services responding to pandemics must develop these novel tests quickly to track the progress of a new epidemic or pandemic strain. Virus isolation is still required for vaccine design, a process that is coordinated nationally by public health services and internationally by the WHO.

Treatment, prevention and control

Treatment is usually symptomatic; secondary bacterial infections require appropriate antibiotics. Inactivated viral vaccines are prepared from the currently circulating viruses each year. Vaccination provides 70% protection and is recommended for individuals at risk of severe disease, such as those with cardiopulmonary disease or asthma. Influenza can be treated with the neuraminidase inhibitors zanamivir and oseltamivir, which shorten the duration of symptoms. They are indicated for patients who are at risk of severe complications and may have value in slowing the progression of a pandemic and reducing the associated mortality. Recent developments utilizing molecular cloning techniques have shortened the time taken to produce novel vaccines in response to a pandemic, which proved useful in the swine 'flu pandemic. Research continues to find a vaccine antigen that is effective but is not variable.

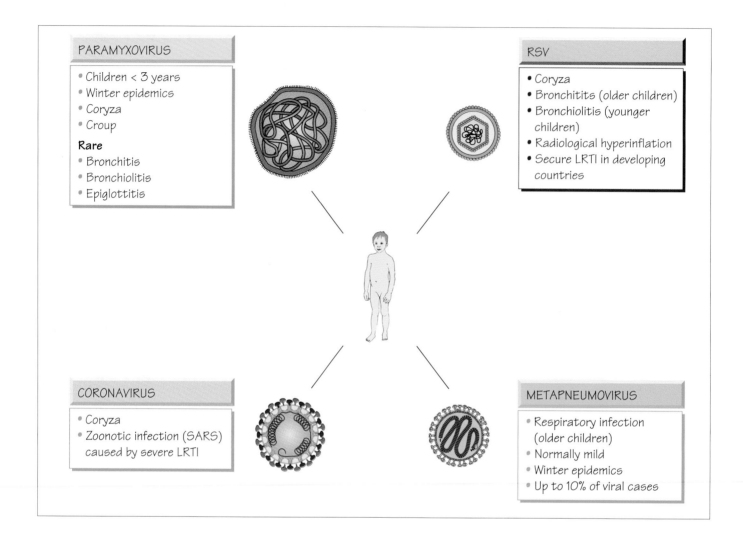

PARAMYXOVIRUS

- Children < 3 years
- Winter epidemics
- Coryza
- Croup

Rare
- Bronchitis
- Bronchiolitis
- Epiglottitis

RSV

- Coryza
- Bronchitits (older children)
- Bronchiolitis (younger children)
- Radiological hyperinflation
- Secure LRTI in developing countries

CORONAVIRUS

- Coryza
- Zoonotic infection (SARS) caused by severe LRTI

METAPNEUMOVIRUS

- Respiratory infection (older children)
- Normally mild
- Winter epidemics
- Up to 10% of viral cases

Parainfluenza virus

This is a fragile, enveloped paramyxovirus (150–300 nm) containing a single strand of negative-sense RNA (15 kb). It has four types that share antigenic determinants.

Pathogenesis and epidemiology

The virus attaches to host cells, where the envelope fuses with the host cell membrane. The virus multiplies throughout the tracheo-bronchial tree. Infection, which is transmitted by the respiratory route, peaks in the winter, with the highest attack rates occurring in children under 3 years old.

Clinical features

In this common, self-limiting condition, which usually lasts 4–5 days, children are distressed, coryzal and febrile. In young children, hoarse coughing often alternates with hoarse crying and is associated with inspiratory stridor secondary to laryngeal obstruc-

tion (croup). Rarely, bronchiolitis, bronchopneumonia or acute epiglottitis may develop, signalled by reduced air entry and cyanosis.

Diagnosis and treatment

Diagnosis is clinical. Direct immunofluorescence gives rapid results; viral isolation and reverse transcriptase nucleic acid amplification tests (NAATs) are available as part of a respiratory virus screen. Treatment is symptomatic (e.g. paracetamol and humidification). Severe infection can be treated with ribavirin and humidified oxygen.

Respiratory syncytial virus

This enveloped paramyxovirus (120–300 nm) containing a single strand of negative-sense RNA attaches to host cells by 12-nm glycoprotein spikes. There is antigenic variation within the two types, designated A and B.

Medical Microbiology and Infection at a Glance, Fourth Edition. Stephen H. Gillespie, Kathleen B. Bamford. © 2012 John Wiley & Sons, Ltd.

Epidemiology

Respiratory syncytial virus (RSV) is found worldwide, infecting children during the first 3 years of life. There are yearly epidemics in the winter months in temperate countries and in the rainy season in tropical countries. RSV spreads readily in the hospital environment. Patients who are elderly and frail, and those with a compromised respiratory tract can develop serious infection.

Clinical features

Coryza develops after a 4 to 5-day incubation period. In 40% of cases bronchitis develops in older children and bronchiolitis in the very young. Severe disease can develop quickly but, with intensive care, mortality is very low. Children with bronchiolitis are febrile and tachypnoeic, with chest hyperinflation, wheezing and crepitations. Cyanosis is rare. The radiological appearances are variable and include hyperinflation and increased peribronchial markings.

Diagnosis and treatment

Direct immunofluorescence or enzyme immunoassay (EIA) of nasopharyngeal secretions is rapid. Many laboratories use reverse transcriptase NAAT for diagnosis. The virus can be cultivated.

Treatment for RSV infection is based on symptomatic relief and humidification. Severe cases may require hospitalization and humidified oxygen. Severely ill, immunocompromised patients may benefit from aerosolized ribavirin.

Prevention

There is no currently available vaccine.

Coronavirus

This is a spherical enveloped virus (80–160 nm) with positive-sense linear single-stranded RNA (27 kb); the envelope contains widely spaced club-shaped spikes. Coronaviruses cause a coryza-like illness similar to that of rhinovirus. The virus has been observed in the faeces of patients with diarrhoeal disease and asymptomatic subjects. Diagnosis is by serology using a complement fixation test (CFT) or EIA, by detection of coronavirus-specific antigens or by electron microscopy.

A coronavirus that emerged in China was associated with severe pneumonia (SARS). It was transmitted by the respiratory and oral route; mortality was approximately 10%, but higher in elderly people and patients who were immunocompromised. Healthcare workers were vulnerable to infection, so stringent precautions were required to prevent hospital transmission. Coordinated infection control has permitted eradication of the virus.

Metapneumovirus

Human metapneumovirus, a paramyxovirus, has recently been identified from children with acute respiratory tract infections. It accounts for just under 10% of cases that occur in the winter months, causing a clinical syndrome that is similar to RSV infection. Dual infection with RSV is associated with severe disease. Diagnosis is by reverse transcriptase NAAT.

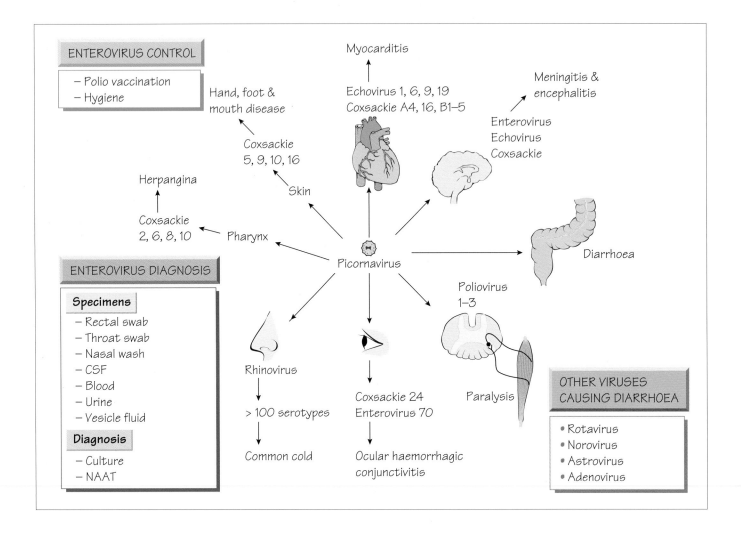

ENTEROVIRUS CONTROL

– Polio vaccination
– Hygiene

Hand, foot & mouth disease

Myocarditis

Echovirus 1, 6, 9, 19
Coxsackie A4, 16, B1–5

Meningitis & encephalitis

Enterovirus
Echovirus
Coxsackie

Coxsackie 5, 9, 10, 16

Skin

Herpangina

Coxsackie 2, 6, 8, 10

Pharynx

Picornavirus

Diarrhoea

Poliovirus 1–3

ENTEROVIRUS DIAGNOSIS

Specimens

– Rectal swab
– Throat swab
– Nasal wash
– CSF
– Blood
– Urine
– Vesicle fluid

Diagnosis

– Culture
– NAAT

Rhinovirus

> 100 serotypes

Common cold

Coxsackie 24
Enterovirus 70

Ocular haemorrhagic conjunctivitis

Paralysis

OTHER VIRUSES CAUSING DIARRHOEA

• Rotavirus
• Norovirus
• Astrovirus
• Adenovirus

Enterovirus

Enteroviruses are picornaviridae with three serotypic groups: poliovirus, coxsackievirus and enteric cytopathic human orphan (ECHO) virus (echovirus). Later isolates have been designated with a number (e.g. enterovirus 68–72).

Enteroviruses are unenveloped, icosahedral, positive-sense RNA viruses that encode for four proteins.

Pathogenesis

• The virus enters cells by a specific receptor that differs for different virus types, therefore defining tissue tropism.
• The virus is usually acquired via the intestinal tract, causing subsequent viraemia and invasion of reticuloendothelial cells.
• Secondary viraemia leads to invasion of target organs (e.g. meninges, spinal cord, brain or myocardium).
• Poliovirus appears to spread along nerve fibres; if significant multiplication occurs within the dorsal root ganglia, the nerve fibre may die, with resultant motor paralysis.

Epidemiology

• Enteroviruses are spread by the faecal–oral route.
• In developing countries infection occurs early in life; it occurs later in industrialized countries.
• Infection can occur in parents and carers of infants who have received the live vaccine.

Clinical features

Polio may present as a minor illness (abortive polio), as aseptic meningitis (non-paralytic polio), with lower motor neurone damage and paralysis (paralytic polio), or as a late recrudescence of muscle wasting that occurs sometimes decades after the initial paralytic polio (progressive postpoliomyelitis muscle atrophy). In paralytic polio, muscle involvement is maximal within a few days after commencement of the paralysis; recovery may occur within 6 months.
• Aseptic meningitis (see Chapter 49) and, rarely, severe focal encephalitis or general infection may present in neonates.

Medical Microbiology and Infection at a Glance, Fourth Edition. Stephen H. Gillespie, Kathleen B. Bamford. © 2012 John Wiley & Sons, Ltd.

- Herpangina, a self-limiting, painful, vesicular pharyngeal infection, is caused by some types of coxsackievirus.
- Coxsackie B causes acute myocarditis (see Chapter 48).
- Hand, foot and mouth disease is characterized by a vesicular rash of the palms, mouth and soles that heals without crusting.

Diagnosis and treatment
- Diagnosis is usually by nucleic acid amplification test (NAAT) of CSF, throat swab and faecal specimen.
- Culture is available.
- The multiplicity of serotypes makes serological diagnosis impractical.
- Treatment is supportive care but pleconaril shows benefit in the treatment of enteroviral meningitis. Artificial ventilation may be required in the case of polio.

Prevention
Two vaccines are available: the oral live attenuated Sabin and the killed parenteral Salk vaccine. Now that polio is limited to a few countries, the inactivated poliovirus vaccine (IPV) is used.

Rhinovirus
- Rhinovirus is responsible for the common cold.
- More than 100 serological types exist.
- It has a short incubation period (2–4 days).
- The virus is excreted whilst symptoms are present.
- Transmission is by contact.

The virus infects the upper respiratory tract, invading only the mucosa and submucosa. The primary symptoms of headache, nasal discharge, upper respiratory tract inflammation and fever may be followed by secondary bacterial infections such as otitis media and sinusitis. Infection occurs worldwide with a peak incidence occurring in the autumn and winter. Immunity after infection is poor because of the multiplicity of serotypes. Ruprintrivir given by nasal spray has been shown to shorten symptoms in clinical trials. A vaccine is impractical.

Rotavirus
Rotaviruses are unenveloped viruses that contain 11 double-stranded RNA segments coding for nine structural proteins and several core proteins.

Pathogenesis
Rotaviruses infect small-intestinal enterocytes; damaged cells are sloughed into the lumen, releasing viruses. Diarrhoea is caused by poor sodium and glucose absorption by the immature cells that replace the damaged enterocytes.

Epidemiology
Rotaviruses are the main cause of viral diarrhoea, occurring usually in children between 6 months and 2 years of age. Morbidity is highest in the young in developing countries. There are seasonal peaks in the winter in temperate countries. Antibody to the virus does not confer immunity to further infection.

Diagnosis
- Diagnosis by reverse transcriptase NAAT is most sensitive.
- Antigen can be detected by enzyme immunoassay (EIA).
- The virus can be visualized by electron microscopy (EM).

Treatment and prevention
Treatment is symptomatic and supportive. The risk of infection can be reduced by provision of adequate sanitation. Vaccines have been introduced into countries where rotavirus morbidity and mortality are high.

Norovirus and astrovirus
Noroviruses are caliciviruses that cause outbreaks of acute diarrhoea and vomiting in hospitals and care homes, on cruise liners and in other confined communities. Infection is transmitted by the faecal–oral and aerosol routes with symptoms developing after a short incubation period (24–48 h). The viruses can be divided into five genogroups. Astroviruses are small spherical particles; more than five serotypes have been recognized.

Virus replication occurs in the mucosal epithelium of the small intestine, which results in broadening and flattening of the villi and hyperplasia of crypt cells.
- Infection usually causes a self-limiting, acute diarrhoeal illness.
- It can present with sudden-onset, projectile vomiting and explosive diarrhoea.
- Sudden outbreaks of norovirus infection may occur in institutions, requiring the units to close to new admissions.
- Diagnosis is made by NAAT.
- Sequencing is required for epidemiological purposes and to monitor the design of future NAAT detection assays.
- Prevention is by isolation, ward closure and good hand-washing technique.

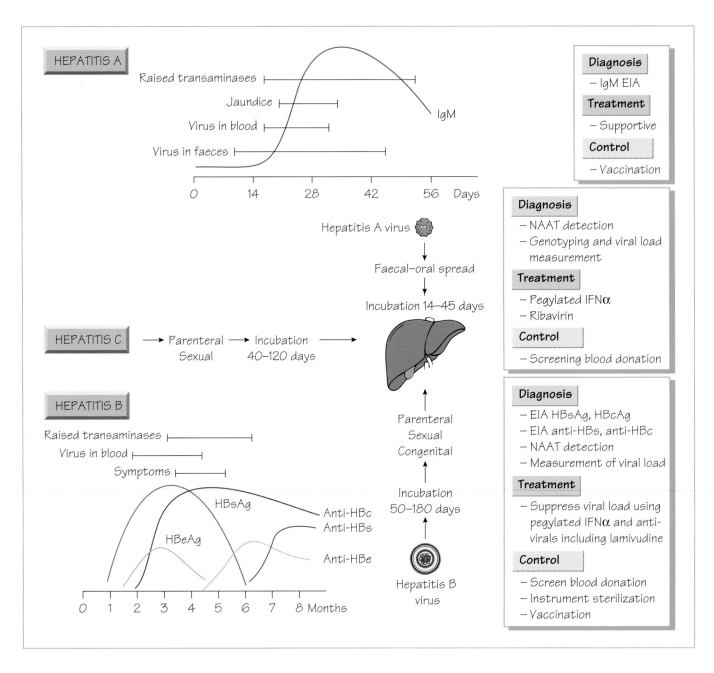

Hepatitis A

Hepatitis A virus (HAV) is a *Hepatovirus* related to the Enteroviruses (see Chapter 36) with four genotypes.

Transmission is by the faecal–oral route. Institutional outbreaks are associated with summer and point-source outbreaks follow faecal contamination of water or food (e.g. oysters). Seroprevalence is highest in individuals of lower socioeconomic groups. Anicteric infection is more common in the young; the risk of symptomatic disease increasing with age. Infection is characterized by a 'flu-like illness followed by jaundice, with most patients making an uneventful recovery. Virus is shed in stool before jaundice appears.

Diagnosis

• Anti-HAV IgM is diagnostic appearing before jaundice develops and persisting for 3 months.
• IgG antibodies determine a patient's immune status.
• HAV RNA can be detected in the blood and stool during the acute phase of infection by nucleic acid amplification test (NAAT).

Treatment and prevention

Treatment is symptomatic and chronic hepatitis does not occur. Adequate sanitation and good personal hygiene will reduce the transmission of HAV. Vaccination against HAV is recommended for travellers to high-risk areas, patients with chronic liver infection and individuals with high-risk occupations (e.g. healthcare workers, sewage workers). Passive immunity can be provided using human immunoglobulin.

Hepatitis B

Hepatitis B (HBV), a hepadnavirus, is an enveloped virus that contains partially double-stranded DNA encoding surface antigen (HBsAg), core antigen (HBcAg), pre-core protein (HBeAg), a large active polymerase protein and transactivator protein. The virus replicates through a reverse transcriptase. HBV is transmitted by parenteral, congenital and sexual routes. A quarter of the global population is infected.

Clinical features

• HBV infection has a long incubation period (up to 6 months).
• Acute hepatitis of variable severity develops insidiously.
• Fulminant disease carries a 1–2% mortality and 10% of patients develop chronic hepatitis complicated by cirrhosis or hepatocellular carcinoma.
• Congenital infection carries a high risk of hepatocellular carcinoma.

Diagnosis

• Immunoassays for HBsAg, HBeAg, HBcAg and associated antibodies enable the diagnosis of acute infection and previous exposure (see Figure).
• Viral load can be measured by NAAT and sequencing for resistance mutations allows monitoring of therapy and directs drug choice.

Treatment and prevention

• Pegylated α-interferon.
• Lamivudine, adefovir, entecavir, tenofovir, telbivudine and clevudine have antiviral efficacy. Emtricitabine and valtorcitabine are nearing clinical introduction.
• Therapy should be considered in chronic infection as responders have a reduced risk of liver damage and liver cancer in the long-term. HBeAg seroconversion is often taken as a mark of treatment success.
• Those at high risk should be immunized with recombinant HBV vaccine.
• Vaccine and specific immunoglobulin should be administered to neonates of infected mothers to reduce transmission.
• Blood donations must be effectively screened.
• Needle-exchange programmes for drug misusers and sexual-health education schemes can help to reduce transmission.

Hepatitis C

Hepatitis C (HCV) is a sense RNA virus encoding a single polypeptide. Transmission is mainly through infected blood. Seroprevalence is approximately 1% in healthy blood donors, higher in developing countries and highest in high-risk groups, such as those who have received unscreened transfusions. Healthcare workers are at risk. Sexual transmission and vertical transmission do occur but are uncommon.

Clinical features

Infection may cause a mild acute hepatitis but many cases are asymptomatic; fulminant disease is rare. HCV infection persists in up to 80% of patients; up to 35% of these develop cirrhosis, liver failure and hepatocellular carcinoma between 10 and 30 years later. This occurs because frequent virus mutation results in immunologically distinct 'quasi-species', which allow the organism to escape immunological control.

Diagnosis

• HCV cannot be cultured.
• Diagnosis is by antibody and antigen detection.
• A NAAT is available.
• Sequencing to determine genotype defines the likelihood of response to therapy (see below).
• Treatment is monitored by measurement of viral load.

Treatment and prevention

• Ribavirin and pegylated α-interferon.
• Response is best in patients with genotypes 1 and 2 and those with low initial viral load, but up to 80% will clear the virus.
• Liver fibrosis or necrotic inflammation from HCV infection is an indication for liver transplantation.
• Preventive measures are similar to those employed against HBV.
• There is no vaccine.

Hepatitis D

This defective RNA virus is surrounded by an HBsAg envelope and is transmitted with and in the same way as hepatitis B virus or as a super-infection in an HBV carrier. Although asymptomatic infection may occur, hepatitis D (HDV) is associated with severe hepatitis and an accelerated progression to carcinoma. A real-time NAAT is the most rapid method of making the diagnosis but antigen detection or IgM antibody detection by enzyme immunoassay (EIA) can also provide confirmation. Preventive measures for HBV also protect against HDV.

Hepatitis E

• Hepatitis E is a small, single-strand, non-enveloped RNA virus.
• Transmission is by the faecal–oral route.
• Outbreaks occur after contamination of water supplies or food.
• It is found in Asia, Africa and Central America.
• It usually causes a self-limiting hepatitis of varying severity.
• Diagnosis is by IgM or NAAT.
• Infection is prevented by hygiene measures.

Viral hepatitis can also be caused by other viruses (e.g. cytomegalovirus [CMV], herpes simplex and Epstein–Barr virus [EBV]).

38 Tropical, exotic or arbovirus infections

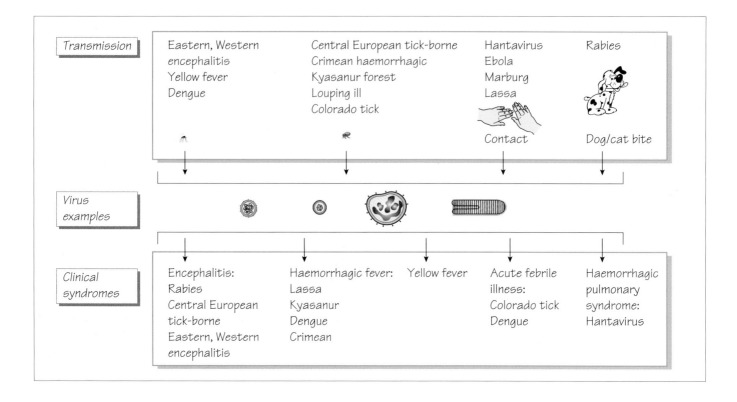

More than 100 viruses can cause encephalitis or haemorrhagic fever. Almost all are zoonoses, where the human is an accidental host that has come into contact with the natural life cycle. They are transmitted by direct contact with blood and body fluids or by the bite of arthropods, such as mosquitoes, ticks and sandflies. Some infections are associated with a high mortality.

Rabies

Rabies is a rhabdovirus infection that, once symptoms develop, causes a fatal encephalomyelitis.
• It is a bullet-shaped, negative-sense RNA enveloped virus.
• It infects warm-blooded animals worldwide.
• The virus is found in saliva and is transmitted to humans through the bite of an infected animal.
• Two epidemiological patterns exist: urban rabies, which is transmitted by feral and domestic dogs; and sylvatic rabies, which is endemic in small carnivores in the countryside. Dog-bites are responsible for most infections.
• Bats, raccoons and skunks are an important reservoir and vector of infection in the Americas; the red fox is the reservoir of infection in Europe.
• The virus enters via the motor endplates, spreading up the axons to enter the brain. Sites with short neural connections to the central nervous system have the shortest incubation period (7 days), whereas a bite on the foot may have an incubation period of 100 days.

• Bite depth and viral inoculum also influence the incubation period.
  A prodromal fever, nausea and vomiting precede disease, which takes one of two forms: furious rabies (hyperexcitability, hyperreactivity, hydrophobia) or dumb rabies (an ascending paralysis). Disease is progressive and inevitably fatal. Diagnosis is based on the clinical and epidemiological features, confirmed by specific fluorescence in corneal scrapings, by brain biopsy, or by the finding specific rabies antibody.
  The disease may be prevented by pre-exposure vaccination, wound care, local antiserum, systemic hyperimmunoglobulin and a postexposure vaccination course with the human diploid cell vaccine. Pre-exposure vaccination is reserved for those in a high-risk group (e.g. vets and travellers to remote regions of endemic countries).

Yellow fever

Yellow fever virus is a flavivirus, an enveloped positive-sense RNA virus, transmitted by *Aedes aegypti*. Yellow fever is a zoonosis in which humans are an accidental host (sylvatic disease), but an urban cycle results in periodic human epidemics.

Clinical features
• Infection may be asymptomatic or may cause acute hepatitis and death from hepatic necrosis.
• A short incubation period is followed by fever, nausea and vomiting, and later by jaundice.

Medical Microbiology and Infection at a Glance, Fourth Edition. Stephen H. Gillespie, Kathleen B. Bamford. © 2012 John Wiley & Sons, Ltd.
Published 2012 by John Wiley & Sons, Ltd.

- Haemorrhagic manifestations may develop and vomitus may be black because of digested blood (vomito negro).
- The mortality rate is high, but patients who recover do so completely.

Diagnosis is clinical, supported by nucleic acid amplification test (NAAT), culture and serology. Disease prevention is by mosquito control and vaccination with the live attenuated vaccine.

Dengue
- Dengue is a flavivirus related to yellow fever virus with four serotypes.
- It is transmitted by *Aedes* mosquitos.
- The incubation period is 2–15 days.
- It is found throughout the tropics and the Middle East.
- Epidemics occur when a new serotype enters the community or a large number of susceptible individuals move into an endemic area. Urban epidemics can be explosive and severe.
- Common features include a sudden onset of fever and chills, and headache with pains in the bones and joints. The fever may be biphasic and a mild rash may also be present.
- Dengue haemorrhagic syndrome causes severe shock and bleeding with mortality of 5–10%.
- Diagnosis is by NAAT, serology or culture.
- Prevention is by mosquito control.
- Treatment is symptomatic.

Japanese B encephalitis
- This is a mosquito-borne flavivirus infection that causes encephalitis with a high mortality.
- The natural reservoir is in pigs.
- It causes abrupt-onset fever and severe headache, nausea and vomiting. Convulsions can occur.
- There may be permanent cranial nerve or pyramidal tract damage.
- Prevention is by vaccination.

West Nile virus
This is a flavivirus infection from Africa, which has been found in North America since 1999 and has spread across the continent into Canada, Latin America and the Caribbean, that causes an encephalitis-like syndrome.

Lassa fever
- Lassa fever is a severe haemorrhagic fever caused by an arenavirus.

- It is endemic in West Africa.
- It is transmitted from house rats to humans and from person to person by contact.
- Patients present with fever, mouth ulcers, myalgia and haemorrhagic rash.
- Diagnosis is clinical and depends on exposure history.
- Confirmation is by NAAT or serology.
- Ribavirin improves outcome if given early and may be given as postexposure prophylaxis to contacts.
- Special isolation is required in hospital.

Ebola and Marburg virus
- These viruses are found in Africa and are transmitted to humans from primates or from a rodent reservoir.
- They cause haemorrhagic disease with high fever and mortality.
- They may be transmitted in the hospital environment.
- Treatment is supportive and with hyperimmune serum.
- Control is not possible as the reservoir is not confirmed.
- Special isolation is required in hospital.
- A vaccine using vesicular stomatitis virus encoding Ebola and Marburg antigens is in development.

Hantavirus
This bunyavirus infection is transmitted to humans from rodents and causes either a haemorrhagic fever with renal failure or hantavirus pulmonary syndrome. The disease occurs widely throughout the world. Person-to-person spread does not appear to take place. The incubation period is 2–3 weeks, followed by fever, headache, backache and injected conjunctiva and palate. Hypotension, shock and oliguric renal failure follow. The mortality rate is about 5%.

Diagnosis is based on NAAT, serology and culture.

Nipah and Hendra virus
Nipah virus, a paramyxovirus, causes severe disease in humans and animals. It is found in South Asia and causes febrile encephalitis with a high mortality rate. The reservoir is probably fruit bats, with human infection from contact with bats or an intermediate animal host such as pigs. Person-to-person spread occurs. The related, rarer Hendra virus is also acquired from bats and causes an influenza-like syndrome or encephalitis.

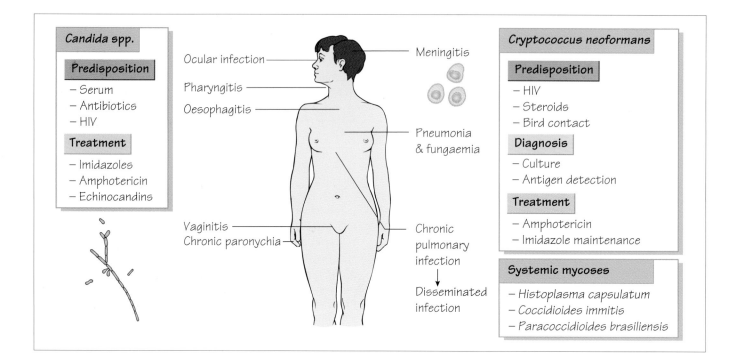

Fungi cause a wide range of diseases, ranging from cutaneous dermatophyte infections to invasive infection in the severely immunocompromised patient. They may have a yeast-like morphology (see below), or be filamentous (see Chapter 40).

Candida spp.

Candida spp. are widely distributed in the environment. They form part of the normal commensal population of the skin, gastrointestinal tract and female genital tract. Following the use of broad-spectrum antibacterial agents, fungal overgrowth may develop into infection. Patients with immunodeficiencies are particularly susceptible to this progression. Most infections are caused by *Candida albicans*. Infection with other species such as *C. tropicalis*, *C. parapsilosis*, *C. glabrata* and *C. pseudotropicalis* are a problem in immunocompromised patients because they may be resistant to the antifungal agents used in therapy or prophylaxis.

Pathogenesis

Although these organisms possess melanin, adhesins and extracellular lipases and proteinases, they have only modest capacity to invade. Infection occurs when the natural resistance provided by the normal bacterial flora is altered by antibiotics, or where there is a severe loss of immune function.

Clinical features

Candida spp. cause pain and itching with creamy curd-like plaques on mucosal surfaces that bleed when removed. Skin and nail-bed infections are common. In immunocompromised patients, phar-yngitis and oesophagitis can be severe; the associated dysphagia may lead to weight loss and is an AIDS-defining illness. Systemic invasion is common in neutropenic patients. *Candida* spp. may also cause systemic and line-associated infection following broad-spectrum antimicrobial therapy in intensive care patients.

Laboratory diagnosis

Diagnosis is by microscopy, culture or nucleic acid amplification test (NAAT). The significance of each isolate is determined in relation to the overall clinical picture. Species identification is by biochemical testing or increasingly by sequencing the 18S rRNA gene.

Antifungal susceptibility

Candida spp. are susceptible to amphotericin, with the exception of *C. lusitaniae*. They are usually susceptible to the imidazoles (e.g. fluconazole) and to 5-flucytosine.

Cryptococcus neoformans

Cryptococcus neoformans is a saprophyte and animal commensal; the composition of pigeon faeces favours its growth. It is a rare cause of chronic lymphocytic meningitis in patients with lymphoma, those taking steroid or cytotoxic therapy and those with intense exposure, such as pigeon fanciers. *Cryptococcus* is an important pathogen in patients with T-cell deficiency.

Pathogenesis

The pathogenicity depends on an antiphagocytic capsule, melanin production and several lytic enzymes.

Clinical features

Infection usually presents as subacute meningitis, although pneumonia and fungaemic shock are recognized. In patients with AIDS, relapses are common and lifelong suppressive therapy is necessary.

Laboratory diagnosis

Infection is diagnosed by microscopy in CSF using Gram stain or India ink, or by detection of the capsular polysaccharide antigen by latex-agglutination. It can be cultured and identified by biochemical tests or 18S rRNA sequencing.

Treatment

Liposomal amphotericin is the treatment of choice and flucytosine and fluconazole may also be used.

Pityriasis versicolor

Malassezia furfur infects the stratum corneum, causing brown, scaly macules. Patients with AIDS may develop severe dermatitis. Topical application of antifungal agents is usually successful.

Systemic yeast infections

Five main species of yeast are associated with systemic infection: *Histoplasma capsulatum*, *H. capsulatum* var. *duboisii*, *Blastomyces dermatitidis*, *Coccidioides immitis* and *Paracoccidioides brasiliensis*.

These organisms have a defined geographical distribution: south-west USA, South America and Africa. Infection is acquired by the respiratory route. Severe disease is more likely in patients with reduced cell-mediated immunity.

Clinical features

Although usually asymptomatic or self-limiting, pulmonary or cutaneous infection may disseminate in infants or immunocompromised patients, causing severe illness.

Laboratory diagnosis

These infections are diagnosed by microscopy and culture of blood, sputum, CSF, urine or pus. The organisms are hazardous and should be handled in a specialized containment facility.

Treatment

Patients with severe disease may be treated with amphotericin B.

Antifungal compounds

Azoles

The azole group of compounds (clotrimazole, miconazole, fluconazole and itraconazole) act by blocking the action of cytochrome P450 and sterol 14α-demethylase. This latter enzyme allows the incorporation of 14-methyl sterols into the fungal membrane, instead of ergosterol. Resistance can develop during long-term treatment.

Clotrimazole and miconazole are frequently used as topical preparations for minor infections.

Fluconazole

Fluconazole can be given orally, topically and parenterally. It is widely distributed, crosses the blood–brain barrier and is active against *Candida* and *Cryptococcus* but not against filamentous fungi. It is used for the prophylaxis and treatment of cryptococcal infections and treatment of superficial and systemic candidiasis. Although well tolerated, it may cause liver enzyme abnormalities and has significant drug interactions, increasing the serum concentration of phenytoin, ciclosporin and oral hypoglycaemic agents and reducing the rate of warfarin metabolism.

Itraconazole

In addition to being effective against *Candida* spp., *C. neoformans* and *Histoplasma*, itraconazole also displays activity against filamentous fungi, including *Aspergillus* and the dermatophytes. It is indicated in the treatment of invasive candidiasis, cryptococcosis, aspergillosis, superficial mycoses and pityriasis versicolor. Resistance is rare. It is well absorbed and can be given orally, achieving high tissue concentrations.

Voriconazole and posoconazole

Voriconazole is a broad-spectrum triazole that is active against many yeasts and moulds including *Aspergillus*. It has been reported to have a better success rate in proven invasive *Aspergillus* infection than amphotericin, but treatment is associated with transient visual disturbance. Posoconazole has a wide spectrum of activity. Further agents are in development.

Flucytosine

This synthetic fluorinated pyrimidine inhibits *Candida* spp., *C. neoformans* and some moulds. The drug disrupts protein synthesis. It is well absorbed orally and can be given intravenously. Adverse events include bone marrow suppression, thrombocytopenia and abnormal liver function tests. Resistance develops rapidly with monotherapy.

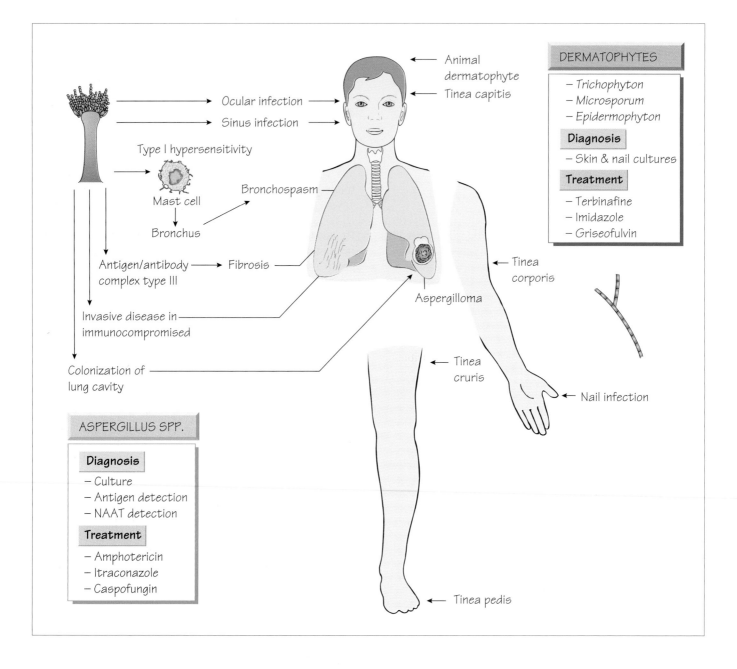

Aspergillus spp.

Aspergillus spp. are ubiquitous, free-living, saprophytic organisms; *A. fumigatus*, *A. niger*, *A. flavus* and *A. terreus* are associated with human infection.

Pathology and Clinical features

Inhalation of *Aspergillus* spores can provoke a type III hypersensitivity reaction with fever, dyspnoea and progressive lung fibrosis (farmer's lung). Colonization by *Aspergillus* can provoke a type I hypersensitivity reaction that results in intermittent airway obstruction (bronchopulmonary allergic aspergillosis).

Healed tuberculosis cavities or bronchiectasis can become colonized with *Aspergillus*, creating a 'fungus ball' or aspergilloma. In neutropenic patients, *Aspergillus* infection typically begins in the lungs and may be followed by fatal disseminated disease. Rarely the paranasal sinuses, skin, central nervous system and eye may become infected, often with a poor prognosis.

Laboratory diagnosis

Sputum culture is of limited value. An isolate from bronchoalveolar lavage is diagnostic (98% specificity) but lacks sensitivity.

Medical Microbiology and Infection at a Glance, Fourth Edition. Stephen H. Gillespie, Kathleen B. Bamford. © 2012 John Wiley & Sons, Ltd.

Antibody detection may confirm bronchopulmonary aspergillosis and farmer's lung, but immunocompromised patients rarely produce an antibody response. Enzyme immunoassay (EIA) to detect galactomannan in serial samples is useful. Nucleic acid amplification tests (NAATs) are a useful adjunct to diagnosis.

Treatment and prevention

Bronchopulmonary aspergillosis requires treatment of the airway obstruction with bronchodilators and steroids. Invasive aspergillosis requires treatment with amphotericin B. Itraconazole has activity against *Aspergillus*; voriconazole improves outcome in pulmonary aspergillosis. Surgery may be beneficial in some cases of pulmonary infection. Patients with farmer's lung should avoid further exposure. Neutropenic patients should be managed in rooms with filtered air and infection should be aggressively treated (see Chapter 41).

Other infections

Mucor, *Rhizopus* and *Absidia* may infect severely immunocompromised patients. Sinus infection may spread to the eyes and brain. Pulmonary disease may be complicated by dissemination. These infections are difficult to treat and have a poor prognosis.

Dermatophytes

Three genera of filamentous fungi are implicated in dermatophytosis: *Epidermophyton*, *Microsporum* and *Trichophyton*. Dermatophytes are also grouped according to their reservoir and host preference: anthrophilic (mainly human pathogens); zoophilic (mainly infect animals); and geophilic (found in soil and able to infect animals or humans). Anthrophilic species spread by close contact (e.g. within families, enclosed communities). Transmission of geophilic species is rare. Close contact with animals may give rise to zoophilic infection (e.g. pet owners, farmers and vets).

Clinical features

Dermatophyte infection in the form of ringworm presents as itchy, red, scaly, patch-like lesions that spread outwards leaving a pale, healed centre; chronic nail infection produces discoloration and thickening; scalp infection is often associated with hair loss and scarring. Clinical diagnostic labels are based on the site of infection (e.g. tinea capitis, head and scalp; tinea corporis, trunk lesions).

Lesions are rarely painful, but zoophilic species produce an intense inflammatory reaction with pustular lesions or an inflamed swelling (kerion).

Laboratory diagnosis

Infection of skin and hair by some species may demonstrate a characteristic fluorescence when examined under Wood's light.

Skin scrapings, nail clippings and hair samples should be sent dry to the laboratory. Typical branching hyphal elements may be demonstrated by microscopy in a potassium hydroxide preparation. Dermatophytes take up to 4 weeks at 30 °C to grow on Sabouraud's dextrose agar.

Identification is based on colonial morphology, microscopic appearance (lactophenol blue mount), biochemical tests and sequencing of the 18S rRNA gene.

Treatment

Dermatophyte infections may be treated topically with imidazoles (e.g. miconazole, clotrimazole, tioconazole or amorolfine). Some infections require oral terbinafine for several weeks.

Antifungal compounds

Terbinafine

Terbinafine inhibits squalene epoxidase with resultant accumulation of aberrant and toxic sterols in the cell wall. It is indicated for the oral treatment of superficial dermatophyte infections that have failed to respond to local therapy. Reported adverse effects include Stevens–Johnson syndrome, toxic epidermal necrolysis and hepatic toxicity. Treatment should be continued for up to 6 weeks for skin infections and 3 months or longer for nail infections.

Griseofulvin

Griseofulvin is active only against dermatophytes by inhibition of fungal mitosis. Given orally, it is incorporated into the stratum corneum or nail where it inhibits fungal invasion of new skin and nail. Treatment must be continued until uninfected tissue grows. It is now rarely used.

Polyenes

There are two polyene cyclic macrolides in clinical use, nystatin and amphotericin B, which are active against almost all fungi. Polyenes bind ergosterol in the fungal membrane forming a pore that leads to leakage of the intracellular contents and cell death. Resistance is rare.

Nystatin is used for topical treatment and the prevention of fungal infection in immunocompromised patients. It has no value for the treatment of dermatophyte infections. Amphotericin is given parenterally; liposomal formulations overcome much of the toxicity of earlier compounds enabling higher doses to be given safely.

Echinocandins

The echinocandins act by inhibiting the synthesis of 1,3β-glucan, a homopolysaccharide in the cell wall of many pathogenic fungi. They are active against both *Candida* and *Aspergillus*. New agents in this class, which are now entering clinical use (e.g. caspofungin), are well tolerated and offer a useful alternative for refractory infections.

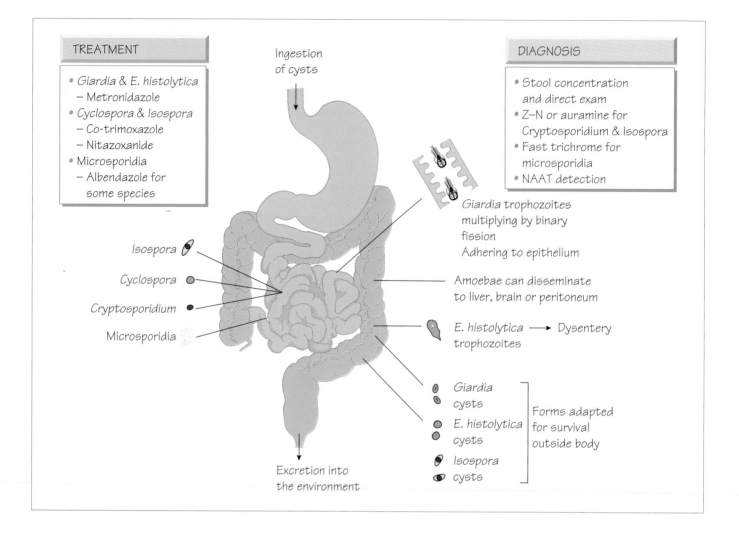

TREATMENT

- *Giardia* & *E. histolytica*
  - Metronidazole
- *Cyclospora* & *Isospora*
  - Co-trimoxazole
  - Nitazoxanide
- Microsporidia
  - Albendazole for some species

Ingestion of cysts

DIAGNOSIS

- Stool concentration and direct exam
- Z–N or auramine for *Cryptosporidium* & *Isospora*
- Fast trichrome for microsporidia
- NAAT detection

Giardia trophozoites multiplying by binary fission
Adhering to epithelium

Isospora

Cyclospora

Cryptosporidium

Microsporidia

Amoebae can disseminate to liver, brain or peritoneum

E. histolytica ⟶ Dysentery
trophozoites

Giardia cysts

E. histolytica cysts

Isospora cysts

Forms adapted for survival outside body

Excretion into the environment

Entamoeba histolytica

Entamoeba histolytica infects the large intestine and is found mainly in developing countries. It is transmitted by the faecal–oral route. The organism causes disease through production of cysteine protease and amoebapore, an epithelial cytotoxin. It is morphologically identical to *E. dispar*, which does not cause disease.

Clinical features

The onset is insidious with little systemic upset: the patient is ambulant but passes frequent small-volume, offensive, bloody stools. Abscesses may develop in the liver or, more rarely, abdomen, lung or brain.

Diagnosis

- Rectal ulceration is seen on sigmoidoscopy.
- Trophozoites are demonstrated in ulcer biopsies.
- Three stool specimens for microscopy, antigen detection and nucleic acid amplification test (NAAT).
- CT and ultrasound for abscesses.

- Serology – may detect liver abscess, but not intestinal infection.

Treatment

- Metronidazole for intestinal infection and abscess.
- Diloxanide furoate or paromomycin kills the chronic cyst stage.

Prevention and control

Steps to ensure that water is boiled and food adequately cooked will reduce the risk of amoebic infection.

Giardia lamblia

Giardia lamblia infection occurs worldwide where poor sanitation allows water supplies or food to be contaminated with cysts from human or animal faeces.

Pathogenesis

Trophozoites multiply in the jejunum and attach to the intestinal wall by a sucking disk. The mechanism for *Giardia* diarrhoea is

Medical Microbiology and Infection at a Glance, Fourth Edition. Stephen H. Gillespie, Kathleen B. Bamford. © 2012 John Wiley & Sons, Ltd.

uncertain and may relate to induced apoptosis. Giardial cysts are excreted in the faeces and survive well in the environment.

Clinical features

These may include:
- bulky, offensive, fatty stools;
- anorexia, crampy abdominal pain, borborygmi and flatus;
- weight loss;
- lactose intolerance or fat malabsorption;
- recurrent attacks of infection in patients with IgA deficiency.

Laboratory diagnosis

- Three concentrated stool samples for microscopy.
- Aspirated jejunal contents can be examined for motile trophozoites.
- Enzyme immunoassay (EIA) and NAAT methods are more sensitive than microscopy.

Treatment

Metronidazole or tinidazole are used. New therapeutic options include albendazole and nitazoxanide. Secondary malabsorption and vitamin deficiency may require investigation and treatment.

Cyclospora cayetanensis

Infection occurs in subtropical and tropical regions from contaminated water supplies; outbreaks from imported soft fruit and fresh herbs have been reported.

Pathogenesis

Cyclospora are found inside vacuoles within the epithelium of the jejunum. There is inflammation, villous atrophy and crypt hyperplasia leading to malabsorption of vitamin B_{12}, folate, fat and D-xylose.

Clinical features

- Infection takes the form of watery diarrhoea.
- A 'flu-like illness and weight loss may also occur.
- Infection is self-limiting, but may last for weeks with continuing fatigue, anorexia and weight loss.
- A prolonged, severe, relapsing disease occurs in individuals who have AIDS.

Diagnosis and treatment

- Microscopy for oocysts in stools.
- NAAT methods are available.
- Co-trimoxazole is an effective treatment, with nitazoxanide as an alternative.

Cryptosporidium

Cryptosporidium parvum is a zoonotic coccidian parasite that is transmitted by milk, water and direct contact with farm animals.

It is naturally resistant to chemical disinfectants, surviving water purification. Person-to-person spread can occur with intimate contact. Infection is common in children and individuals with AIDS. It may interfere with the glucose-stimulated sodium pump in the small intestine, leading to fluid secretion.

Clinical features

- Infection usually produces self-limiting, watery diarrhoea with abdominal cramps.
- Profuse, prolonged diarrhoea in immunocompromised individuals may cause life-threatening fluid and electrolyte imbalance.
- Biliary tree, gallbladder and respiratory tract involvement may occur.

Diagnosis and treatment

- Demonstration of cysts by microscopy.
- Antigen detection or NAAT.
- Nitazoxanide may improve clearance of pathogens but management should aim to reverse immunodeficiency (e.g. treatment of AIDS).

Isospora belli

Closely related to *Cryptosporidium, Isospora belli* presents with a similar clinical picture, usually following tropical travel. Diagnosis is by microscopy of stool. Treatment is with co-trimoxazole; fluoroquinolones or nitazoxanide.

Microsporidia

These are small intracellular protozoa that infect insects, plants and animals. *Enterocytozoon bieneusi, Encephalitozoon cuniculi, Encephalitozoon hellem, Septata intestinalis, Pleistophora* and *Nosema* have been implicated in human infection, usually in immunocompromised patients.

Pathogenesis

Enterocytozoon bieneusi and *S. intestinalis* infect epithelial cells of the small bowel and are associated with diarrhoea. *E. cuniculi* infects macrophages, epithelial cells and vascular endothelial cells in the brain and the kidney plus renal tubular cells. It is associated with hepatitis, peritonitis, diarrhoea, seizures and disseminated infection. Before the advent of HIV infection, microsporidia infection was very rare.

Diagnosis and treatment

Microscopy using fast trichrome, calcofluor white and Ziehl–Neelsen stains can be used. Sensitive NAATs are available to demonstrate organisms.

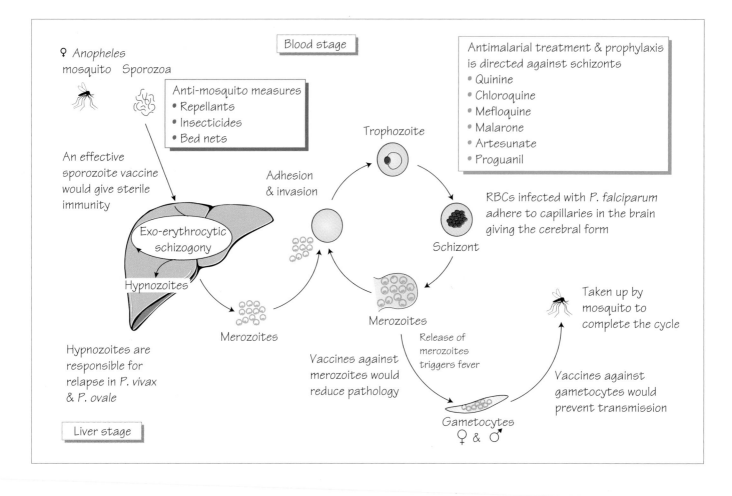

Malaria

Malaria is caused *Plasmodium falciparum, P. vivax, P. ovale* or *P. malariae*. More than 1 million children under the age of 5 years die each year in Africa alone. In the UK there are more than 2000 reported cases and up to 10 deaths every year. Immigrants returning to their home country are at high risk as they have lost their natural immunity and may not take prophylaxis.

Life cycle

Malaria is transmitted by the bite of a female anopheles mosquito that injects sporozoa. The parasites develop in hepatocytes then invade red blood cells and multiply rapidly. They provoke the release of cytokines, which are responsible for many of the signs and symptoms of malaria. Infected red blood cells develop knob-like projections that make them adhere to the capillary wall. This may occur in the brain, causing cerebral malaria.

The life cycle is completed when the sexual stages (gametocytes) are taken up by another mosquito and develop in the mosquito gut into sporozoites, which migrate to the insect salivary glands ready for another bite. *P. vivax* and *P. ovale* develop dormant stages (hypnozoites) that cause relapses.

Clinical features

• Fever or 'flu-like symptoms, although there may be no 'typical' malaria symptoms.
• Residence or travel history.
• Rapid progression is possible in *P. falciparum* infection in the young or travellers.
• Complications of *P. falciparum* may include cerebral malaria, circulatory shock, acute haemolysis and renal failure, hepatitis and pulmonary oedema.
• Other species are associated with milder, more chronic disease.

Diagnosis

• Urgent blood film examination – usually three.
• Antigen detection tests are useful where laboratories have less experience.
• Nucleic acid amplification tests (NAATs) are used to detect drug resistance.

Treatment

• Artesunate combination therapy or quinine is recommended for management of *P. falciparum* malaria.

Medical Microbiology and Infection at a Glance, Fourth Edition. Stephen H. Gillespie, Kathleen B. Bamford. © 2012 John Wiley & Sons, Ltd.

- Intensive care support may be required.
- Chloroquine and primaquine are used for other malaria species.

Prevention and control
- Insecticide-treated bed-nets.
- Insecticide room spraying.
- Covering of exposed skin between dusk and dawn.
- Prophylaxis should be taken according to expert, up-to-date advice.
- Despite extensive research no vaccine is yet available.

Leishmaniasis
This is a protozoan disease transmitted by sandflies, which inject infective promastigotes that adapt to survival in cells of the reticuloendothelial system as amastigotes.

Clinical features
Two disease forms exist: visceral disease (*Leishmania donovani, L. infantum* or *L. chagasi*) and cutaneous disease (*L. major, L. tropica* and *L. aethiopica* in the Old World and *L. braziliensis* and *L. mexicana* in the Americas).
- **Visceral disease:** Fever and wasting are important symptoms. The bone marrow, liver and spleen are infiltrated by parasites so the patient becomes anaemic, leucopenic and thrombocytopenic. Reactive hypergammaglobulinaemia makes patients susceptible to secondary bacterial infections. Untreated patients will deteriorate and die within 2 years.
- **Cutaneous disease:** Chronic, circular skin lesions occur at the site of the bite, with satellite lesions. Some infections with *L. braziliensis* may cause destruction of structures around the mouth and nose (espundia).

Diagnosis and treatment
- Microscopy of skin biopsy, bone marrow sample, blood sample or splenic aspirate and culture.
- NAAT is used for primary diagnosis and speciation.
- Serological diagnosis by direct agglutination and dipstick tests for field use.
- Visceral and cutaneous leishmaniasis can be treated with parenteral liposomal amphotericin B. Alternatives include antimony compounds, paromomycin and oral miltefosine.

Trypanosomiasis
African trypanosomiasis
African trypanosomiasis, caused by *Trypanosoma brucei gambiense* and *T. brucei rhodesiense*, is transmitted by the tsetse fly. Humans are the only host of *T. brucei gambiense*, but antelope or cattle act as the reservoir for *T. brucei rhodesiense*. Parasites in the blood are inhibited by immune responses, but antigenic variation allows the organisms to multiply again. Generalized lymphadenopathy may be present and the skin may appear oedematous. The patient exhibits a hypergammaglobulinaemia and is susceptible to secondary bacterial infection. When parasites invade the brain they cause chronic, progressive encephalitis and the patient lapses into coma. Death is often the result of secondary bacterial pneumonia.

Diagnosis and treatment
Parasites are demonstrated in blood, CSF or lymph-node aspirate. Serological tests are available. Lumbar puncture should only be performed after circulating parasites have been eliminated with primary treatment.

Primary treatment T. brucei gambiense is treated with pentamidine; *T. brucei rhodesiense* with suramin.

Secondary treatment of cerebral infection T. brucei gambiense is treated with melarsoprol or eflornithine +/– nifurtimox; *T. brucei rhodesiense* with melarsoprol.

South American trypanosomiasis
Trypanosoma cruzi, which causes Chagas disease, is transmitted by the bite of reduviid bugs. There are three phases of the disease: acute infection characterized by cutaneous oedema, intermittent fever, shock and a significant mortality in children; latent infection; and late manifestations, such as achalasia, megacolon, cardiac dysrhythmias, cardiomyopathy and neuropathy.

Diagnosis
Parasites are demonstrated by microscopy, or culture in artificial media or laboratory bugs (xenodiagnosis). Serological tests are available.

Treatment
Nifurtimox and benznidazole are useful in the acute phase but efficacy declines as infection progresses. Treatment of complications is mainly palliative (e.g. cardiac pacemakers for heart block secondary to cardiomyopathy, surgery for megacolon).

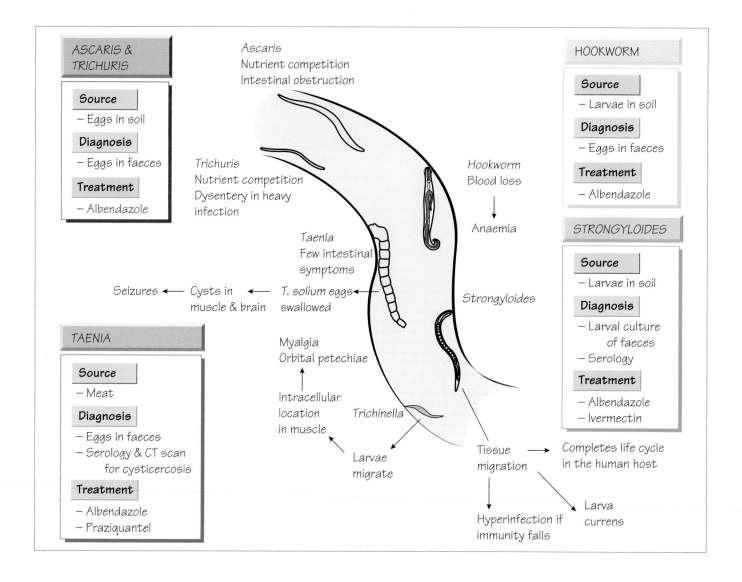

Roundworms and hookworms

Infection by nematodes (roundworms), *Ascaris lumbricoides* and *Trichuris trichiura*, or hookworms, *Necator americanus* and *Ancylostoma duodenale*, is prevalent in developing countries.

Epidemiology

Adult *Ascaris* are found in the intestine; the female worm produces up to 2000 eggs per day. The eggs survive in the soil where they mature into infective forms that can be ingested. After hatching in the small intestine, the organism undergoes a migratory cycle through the liver and lungs, where it is coughed up and swallowed developing into the adult worm in the intestine. Transmission is aided by poor sanitation or when food crops are manured with human faeces. In warm, moist climates, the eggs can survive for many years in the soil. Hookworm eggs hatch into an infective larva that is able to burrow through intact skin to cause infection.

Pathogenesis

These parasites cause disease by competing for nutrients; thus, the severity of symptoms is proportional to the number of worms present (parasite load). Hookworms take blood, leading to iron-deficiency anaemia that can be severe. Heavily infected children have poor growth and lowered school performance, which is attributed to micronutrient deficiency, especially with *Trichuris* infection.

Clinical features

Infection is usually asymptomatic, but heavy *Ascaris* infection may lead to intestinal obstruction and heavy *Trichuris* infection to a dysentery-like syndrome.

Medical Microbiology and Infection at a Glance, Fourth Edition. Stephen H. Gillespie, Kathleen B. Bamford. © 2012 John Wiley & Sons, Ltd.

Diagnosis

The diagnosis is made by examining up to three stool samples for the presence of the characteristic eggs.

Treatment

Intestinal nematodes are treated with albendazole or mebendazole. Improved sanitation is required to control the spread of infection.

Threadworms

Humans are the only host of *Enterobius vermicularis*. Living in the large intestine, the females migrate to the anus where they lay their eggs on the perianal skin. Symptoms are few; thread-like worms may be found in the faeces or patients may complain of perianal itching, often worse at night. Scratching allows contamination of the fingers with larvae-containing eggs which, when placed in the mouth, initiate a new cycle of infection. Occasionally, *Enterobius* can be found in the appendix.

Diagnosis is made by sending an adhesive-tape swab to the laboratory where D-shaped eggs are seen. Treatment is with mebendazole or piperazine. It is often necessary to treat the whole family, repeating treatment after 2 weeks, and hygiene should also be addressed.

Strongyloides stercoralis

Strongyloides stercoralis larvae are passed in the stools, where they either undergo a free-living cycle in the soil or differentiate into infective larvae that invade another host via intact skin. Inside the human host, they can initiate another development cycle. As a consequence, infection with *Strongyloides* can be prolonged. Resistance to *Strongyloides* depends on efficient cell-mediated immunity. Individuals who are infected with human T-cell leukaemia virus 1 (HTLV-1) or are taking steroids are especially susceptible to hyperinfection.

Clinical features

Migrating larvae leave a red, itchy track, which fades after about 48 h. If the patient is given immunosuppressive therapy, uncontrolled multiplication of the *Strongyloides* may occur, which is characterized by fever, shock and the signs of septicaemia and meningitis.

Diagnosis

Stool culture from multiple specimens may reveal infective larvae. Alternatively, samples of jejunal fluid are examined for the presence of larvae. A sensitive enzyme immunoassay (EIA) technique for serum is available.

Treatment

Ivermectin is the optimal drug for treatment, with the imidazoles (e.g. albendazole) as alternatives. Relapse occurs in up to 20% of patients. The hyperinfection syndrome is often accompanied by Gram-negative septicaemia, which requires urgent treatment.

Prevention

The risk of infection can be reduced by wearing appropriate footwear to prevent larvae penetrating the skin.

Tapeworms

Taenia spp.

Two *Taenia* spp. infect humans: the pork worm, *T. solium*, and the beef worm, *T. saginata*. Infection is acquired by eating meat from the intermediate hosts that contains the tissue stages of the parasite.

Pathogenesis and clinical features

Tapeworms compete for nutrients and infections are usually asymptomatic.

Taenia solium can use humans as an intermediate as well as the definitive host. When an individual ingests *T. solium* eggs, the eggs hatch and disseminate, forming multiple cyst-like lesions in the muscles, skin and brain. These 'measly' lesions, similar in appearance to infected pork meat, are known as cysticercosis. Inflammatory responses to parasitic antigens that leak from cysts in the brain may lead to epileptic seizures.

Diagnosis

This is made by finding characteristic eggs in the patient's stool. Cysticercosis is diagnosed by a specific EIA and confirmed by demonstrating the presence of multiple tissue cysts by X-ray, CT or MRI.

Treatment and prevention

Treatment is with praziquantel. Specialist advice should be sought for the management of *Taenia* infections in the central nervous system as severe inflammatory reactions can occur.

Diphyllobothrium latum

Humans are the definitive host of this rare tapeworm, acquiring infection from undercooked freshwater fish. The parasite competes for nutrients and causes deficiency of vitamin B_{12}. The diagnosis is made by detecting the characteristic eggs in faeces and treatment is with praziquantel.

Hymenolepis nana

Humans are the only host of this small tapeworm. Infection is usually asymptomatic and diagnosis is made by detecting the characteristic eggs in the faeces. Treatment is with praziquantel.

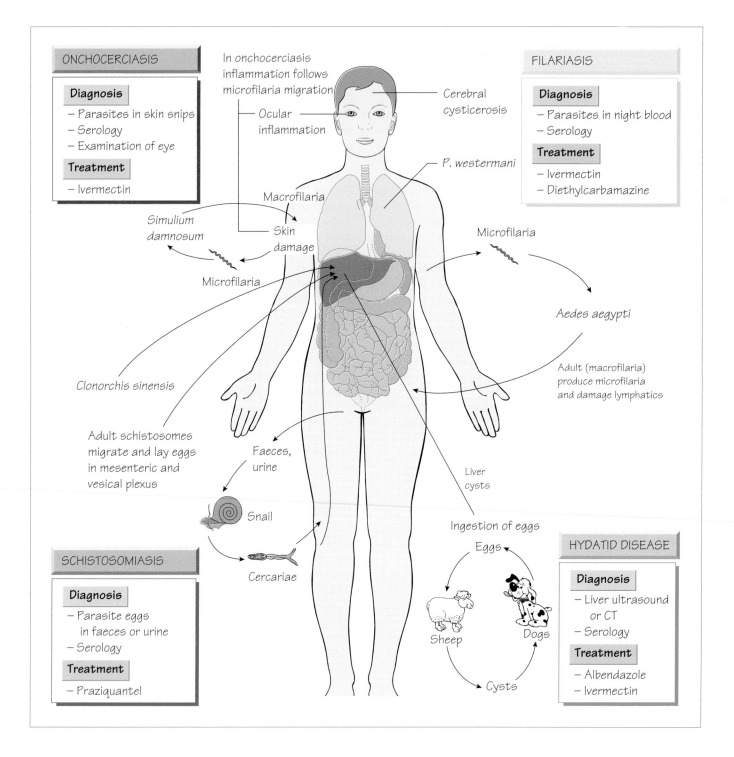

ONCHOCERCIASIS

Diagnosis
- Parasites in skin snips
- Serology
- Examination of eye

Treatment
- Ivermectin

In onchocerciasis inflammation follows microfilaria migration

Ocular inflammation

Cerebral cysticerosis

P. westermani

FILARIASIS

Diagnosis
- Parasites in night blood
- Serology

Treatment
- Ivermectin
- Diethylcarbamazine

Macrofilaria

Simulium damnosum

Skin damage

Microfilaria

Microfilaria

Aedes aegypti

Clonorchis sinensis

Adult (macrofilaria) produce microfilaria and damage lymphatics

Adult schistosomes migrate and lay eggs in mesenteric and vesical plexus

Faeces, urine

Snail

Cercariae

Liver cysts

Ingestion of eggs

Eggs

Sheep

Dogs

Cysts

HYDATID DISEASE

Diagnosis
- Liver ultrasound or CT
- Serology

Treatment
- Albendazole
- Ivermectin

SCHISTOSOMIASIS

Diagnosis
- Parasite eggs in faeces or urine
- Serology

Treatment
- Praziquantel

Medical Microbiology and Infection at a Glance, Fourth Edition. Stephen H. Gillespie, Kathleen B. Bamford. © 2012 John Wiley & Sons, Ltd.
Published 2012 by John Wiley & Sons, Ltd.

Schistosomiasis

Three species infect humans: *Schistosoma mansoni* (Africa and South America); *S. japonicum* (Far East); and *S. haematobium* (Africa). Eggs are excreted in the faeces and urine of infected humans. In areas with poor sanitation, the eggs hatch, releasing a miracidium that invades a snail. After development in the snail, the schistosome cercariae emerge into the environment. They actively penetrate intact skin and develop into male and female adult worms that migrate to the superior, inferior mesenteric or vesical plexus, depending on species, where they lay eggs.

Pathogenesis and clinical features
• **Initial infection:** fever, hepatosplenomegaly, rash and arthralgia.
• **Egg expulsion:** bloody diarrhoea or haematuria.
• **Later:** symptoms and signs are caused by the fibrotic reaction to the eggs in the liver (hepatic fibrosis and portal hypertension), the lungs (pulmonary fibrosis) and bladder. Space-occupying lesions in the brain and spinal cord may lead to seizures.

Diagnosis
Microscopy for eggs in stool, urine, rectal snips or other tissue biopsy. An enzyme immunoassay (EIA) that detects antischistosomal antibody is useful, especially in travellers, and antigen detection methods are available as research tests.

Prevention and control
• Avoidance of contaminated water and the wearing of appropriate clothing when working in the fields.
• Control programmes that target snails.
• Mass treatment can control the disease if sufficient resources are available.

Filariasis

Lymphatic filariasis is caused by *Brugia malayi* and *Wuchereria bancrofti* and transmitted by the mosquito *Aedes aegypti* throughout the tropics. Onchocerciasis is caused by *Onchocerca volvulus* and transmitted by the blackfly, *Simulium damnosum*, in West Africa and South and Central America. Loa (loiasis) is caused by *Loa loa* and transmitted by *Chrysops* flies in West Africa.

Clinical features
Lymphatic filariasis is characterized by acute attacks of fever and lymphoedema, which may be complicated by secondary bacterial infection. After repeated attacks lymphatic vessels are permanently damaged, leading to lymphoedema in the leg, arm or scrotum. Inflammation arises from the response to endobacteria (*Wolbachia* spp., related to *Rickettsia*) that infect the filariae.

Onchocerca adults are located in nodules and microfilariae migrate in the skin, resulting in pruritus and dry, thickened skin. Inflammation in the eye causes blindness.

Loiasis is less damaging and diagnosis is based on fleeting subcutaneous swellings, known as Calabar swellings. Infection may be associated with fever and abnormalities of renal function.

Diagnosis
• **Lymphatic filariasis:** identification of microfilariae in a peripheral blood sample taken at midnight.

• **Onchocerciasis:** microscopic examination of 'pinch biopsies' taken from any affected area, plus the shoulder-blade, buttocks and thighs. If the biopsy is negative, a 50-mg dose of diethylcarbamazine will induce increased itch (the Mazzotti reaction).
• **Loiasis:** *Loa loa* is detected in daytime blood films.
• EIA is also used for diagnosis.

Treatment
Treatment of filarial infection may stimulate an acute inflammatory reaction.
• **Lymphatic filariasis:** diethylcarbamazine or ivermectin together with albendazole.
• **Onchocerciasis:** ivermectin (the addition of tetracycline enhances the effect on microfilarial load by sterilizing the adult worms).

Prevention and control
An international onchocerciasis control programme is under way, using mass treatment of whole populations with ivermectin and doxycycline. Lymphatic filariasis is prevented by mosquito-control measures.

Hydatid disease
See Chapter 54.

Clonorchis sinensis (Opisthorchis sinensis)
Infection is acquired, mainly in the Far East, by eating undercooked fish that contains metacercariae. The adults live in the bile ducts, and eggs are passed in the faeces. Light infections are usually asymptomatic, but heavier infections result in cholangitis and pancreatitis; biliary obstruction and cirrhosis may develop. Cholangiocarcinoma is a late complication. The diagnosis is made by identifying the characteristic eggs in faeces. Patients may be treated with praziquantel. Infection is prevented by adequate cooking of potentially infected fish.

Fasciola hepatica
Humans are accidental hosts of this sheep and cattle parasite. The infective stage is found on freshwater plants such as watercress which, if eaten without cooking, can result in infection. The larvae hatch in the intestine; after maturation and migration, the adults are located in the liver. Patients present with fever and right upper quadrant pain. Low-grade biliary symptoms and liver fibrosis may denote continuing infection. Treatment is with praziquantel.

Paragonimus spp.
Paragonimus spp. infect different organs: *P. westermani*, lungs; and *P. mexicanus*, brain. This rare infection follows the ingestion of undercooked crustaceans. Acute, non-specific symptoms, such as fever, abdominal pain and urticaria, are followed by specific symptoms and signs, such as chest pain, dyspnoea and haemoptysis or central nervous system signs. Diagnosis is made by identification of the characteristic eggs in sputum, imaging, serology or tissue biopsy. Lung fluke is treated with praziquantel; cerebral disease with a combined surgical and medical approach.

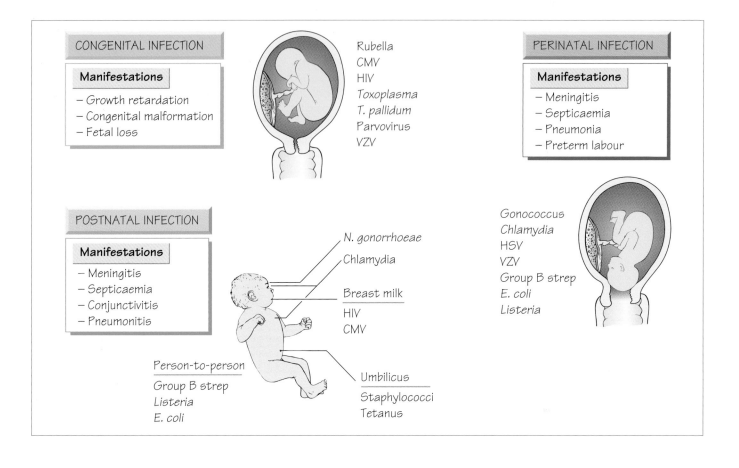

Infection may be acquired across the placenta (intrauterine infection) or contracted during the process of birth or by direct contact with maternal body fluids. Prolonged rupture of the membranes predisposes to fetal infection. Infection can also be transmitted to the neonate after birth from the mother or other contacts.

Congenital rubella

Jaundice associated with hepatitis is often the first sign of congenital rubella. Haemolysis and thrombocytopenic purpura are also common, as is a low-grade meningoencephalitis. Some babies have evidence of metaphyseal dysplasia. Infected infants have low birthweight and fail to attain their expected developmental milestones. There is a high mortality in severely affected infants. Patent ductus arteriosus, cataracts, deafness and retinal pigment dysplasia may be present. Rubella IgM is positive and persists until the third month of life.

The risk to the fetus from maternal infection during the first trimester is more than 60% and some parents will opt for termination of pregnancy. Later the risk is much lower (2% after 20 weeks) and the balance between the chance of fetal damage and the desirability of termination should be considered carefully.

Cytomegalovirus

Infection occurs in less than 1% of births, of which 1% are severely affected. The risk of infection is highest during the first trimester. It presents with prematurity, low birthweight, hepatomegaly, splenomegaly, thrombocytopenia, prolonged jaundice, cerebral irritability, fits and/or abnormal muscle tone or movement.

Microcephaly and sensorineural deafness are the most common problems. Other problems include cerebral calcification, hemiplegia, psychomotor retardation, chorioretinitis and myopathy. Diagnosis depends on demonstrating IgM antibodies or cytomegalovirus (CMV) excretion during the first 20 days of life.

Congenital and intrapartum herpes simplex infections

Primary herpes simplex infections may be accompanied by viraemia when transplacental infection can occur. Infants born with congenital infection tend to have severe disease, with pneumonitis, meningoencephalitis, hepatosplenomegaly and cytopenias. Only a few will demonstrate herpetic skin or mucosal lesions. Treatment with aciclovir reduces mortality from 80–90% to 10–15% and should not wait for laboratory confirmation.

Primary infection may be contracted at birth from maternal genital herpes. Skin, conjunctival, oral or genital lesions develop within a few days, with dissemination in 50% of cases. Treatment is with intravenous aciclovir.

Varicella

Varicella embryopathy follows maternal infection during the first or second trimester of pregnancy; it is transmitted in less than 3% of infected pregnancies. Cicatricial contracture of a limb with hypoplasia, microcephaly or microphthalmia may occur. Non-immune women exposed to chickenpox should be offered postexposure prophylaxis with zoster immune globulin (ZIG) within 10 days of exposure.

Neonatal varicella occurs when the mother develops chickenpox within 1 week of delivery. As neonatal mortality is up to 40%, the neonate should be given ZIG within 48 h of birth if possible and treated with aciclovir if infection develops. Normal immunoglobulin given to the mother will not protect the infant. A vaccine is entering clinical use in some countries.

Listeriosis

Transplacental transmission of *Listeria monocytogenes* occurs during a maternal infection that is often not apparent. Infection in early pregnancy often results in fetal death; later infection is associated with premature labour. Severe bacteraemia, associated with hepatosplenomegaly, meningoencephalitis, thrombocytopenia and pneumonitis, usually complicates neonatal infection. Intrapartum exposure may lead to neonatal infection during the first 2 weeks of life, usually with meningitis and bacteraemia. Blood, CSF, placental tissue and lochia should be cultured. Infected mothers and infants may be a source of infections in the postnatal ward and should be isolated. Ampicillin with or without the addition of gentamicin (for 2–6 weeks) is the treatment of choice.

Syphilis

Congenital infection is now rare as a result of antenatal screening. Affected babies are feverish with features similar to secondary syphilis: rash, condylomata and mucosal fissures. Osteochondritis may cause pain. Persistent rhinitis ('snuffles') is common.

Diagnosis is confirmed by dark-ground microscopy of mucosal or skin lesions. Specific IgM or antibodies persisting after 6 months indicate infection. Late manifestations appear between 12 and 20 years: deafness, optic atrophy or paretic neurosyphilis. Other features include bossing of the frontal bones, chronic tibial periostitis, notching of the incisors, 'mulberry' deformity of the first permanent molar and a high arched palate. The treatment of choice is benzylpenicillin.

Toxoplasmosis

The incidence of toxoplasmosis varies internationally; it is uncommon in the UK, but common in France. Transplacental infection occurs in a third of affected pregnancies. Infection in the first and second trimester is more likely to cause significant fetal disease: the fetus may be stillborn, die soon after birth, or have cerebral calcification, cerebral palsy or epilepsy. Chorioretinitis may not be evident until after birth and may be the only feature. Maternal toxoplasmosis is confirmed by specific IgM antibodies or by seroconversion. IgM antibodies may also be demonstrated in affected neonates. Treatment with spiramycin may reduce the risk of transplacental infection but does not affect the outcome of fetal disease.

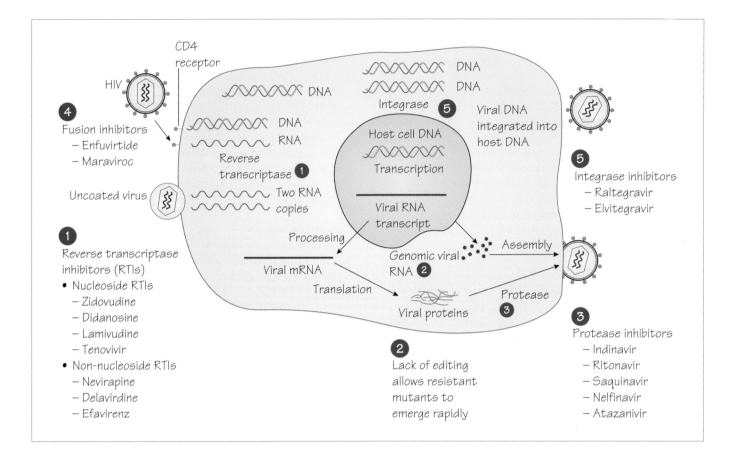

Human immunodeficiency virus

HIV is a spherical, enveloped RNA virus. It is a retrovirus, using reverse transcriptase to produce a DNA copy from viral RNA that is incorporated into the host nucleus to become the template for further viral RNA. Three genes are required for viral replication: *gag*, *pol* and *env*. HIV is classified as a lentivirus. There are two types that are pathogenic for humans: HIV-1, which is most common; and HIV-2, which is found mainly in West Africa and appears to be less virulent.

Epidemiology

Infection with HIV has spread worldwide, transmitted by the parenteral and sexual routes. Infection is most common in individuals at high risk of sexually transmitted diseases, especially those where genital ulceration is common. In developed countries, the main risk groups are intravenous drug users and men who have sex with men; heterosexual transmission is less common but does occur. In developing countries, HIV spreads mainly by heterosexual transmission and through unscreened transfusions or use of contaminated medical equipment. Infection can be transmitted from mother to fetus.

Pathogenesis

The virus principally infects cells with a CD4 receptor (e.g. T cells and macrophages). Viral replication results in progressive T-cell depletion and diminished cell-mediated immunity. Different virus strains display varying affinities for cells that express particular chemokine receptors. Lacking T-cell help, B-cell function is also reduced. HIV causes damage to neural cells and stimulates cytokine release that may also cause neurological damage. Many of the clinical signs of AIDS are caused by secondary infections, which occur when the CD4 count falls.

Clinical features

A few weeks after infection, a mononucleosis-like syndrome may develop with rash, fever and lymphadenopathy. There is a primary viraemia after which the viral load reduces to a steady state, the concentration ('set-point') being related to outcome. The CD4 count declines and, if untreated, reaches a point ($<0.2 \times 10^9$/L) where CD4 function is sufficiently compromised for secondary infections and malignancies to develop, a condition known as acquired immune deficiency syndrome (AIDS). Use of highly active antiretroviral therapy (HAART) delays this development.

Medical Microbiology and Infection at a Glance, Fourth Edition. Stephen H. Gillespie, Kathleen B. Bamford. © 2012 John Wiley & Sons, Ltd.

Bacteria *Mycobacterium* *tuberculosis*, *Mycobacterium avium–intracellulare* (see Chapter 18), *Salmonella*, *Streptococcus pneumoniae*.

Protozoa *Toxoplasma gondii*, *Cryptosporidium parvum*, *Isospora belli*, *microsporidia*.

Fungi *Candida* spp., *Cryptococcus neoformans*, *Pneumocystis carinii*.

Viruses Varicella zoster virus (VZV), human papovavirus.

Malignancy Kaposi sarcoma (HHV-8), non-Hodgkin's lymphoma.

Children with HIV infection are especially vulnerable to childhood viral infections (e.g. measles) and recurrent bacterial infections (e.g. pneumonia).

Diagnostic testing

Diagnosis is by detection of HIV-specific antibody using two different immunoassay methods. As seroconversion may take up to 3 months, an initial negative result should be repeated.

HIV viral load and CD4 count are tested on presentation and regularly during care. Resistance should be tested initially and if a change of therapy is necessary. HLA B5701 and tropism are tested if treatment with abacavir or CCR5 antagonist is contemplated.

Treatment

Antiretroviral drugs are described in Chapter 29. Therapy aims to maintain the virus at fewer than 50 copies, prevent the emergence of resistance, restore immunological function and prevent transmission. Treatment is initiated for any patient with opportunistic infection or when the CD4 count is $<0.350 \times 10^9$/L, although patients with higher counts may benefit. It is also recommended for pregnant women who do not meet these criteria (see Chapter 45). Patients must be aware of the lifelong commitment. Because RNA viruses lack efficient mechanisms for genetic proof-reading, mutations arise rapidly and patients develop drug resistance quickly.

Regimens must be adapted to an individual patient's medical and social condition. Patients infected with resistant virus require tailored regimens. Regimens may include an NNRTI and two NRTI, or a combination that includes a protease or integrase inhibitor. As the immune system starts to recover with treatment, known as **immune reconstitution**, symptoms from opportunistic infection can worsen due to the effects of the enhanced immune reaction.

Prevention
- Avoidance of partners who have a high-risk factor and unprotected intercourse (e.g. by using barrier contraception).
- Screening of blood products.
- Health education and free needle-exchange programmes for intravenous drug users.
- Antigenic diversity has frustrated vaccine development.
- Antiretroviral prophylaxis should be given for infected needle-stick injuries.
- Transmission from mother to child can occur (see Chapter 45).

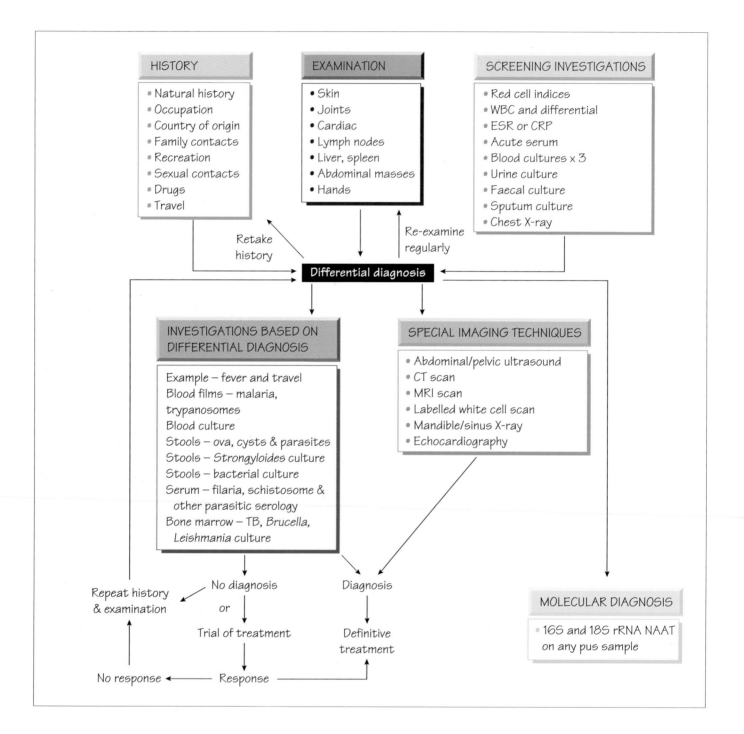

HISTORY
- Natural history
- Occupation
- Country of origin
- Family contacts
- Recreation
- Sexual contacts
- Drugs
- Travel

EXAMINATION
- Skin
- Joints
- Cardiac
- Lymph nodes
- Liver, spleen
- Abdominal masses
- Hands

SCREENING INVESTIGATIONS
- Red cell indices
- WBC and differential
- ESR or CRP
- Acute serum
- Blood cultures x 3
- Urine culture
- Faecal culture
- Sputum culture
- Chest X-ray

Retake history

Re-examine regularly

Differential diagnosis

INVESTIGATIONS BASED ON DIFFERENTIAL DIAGNOSIS

Example – fever and travel
Blood films – malaria, trypanosomes
Blood culture
Stools – ova, cysts & parasites
Stools – *Strongyloides* culture
Stools – bacterial culture
Serum – filaria, schistosome & other parasitic serology
Bone marrow – TB, *Brucella*, *Leishmania* culture

SPECIAL IMAGING TECHNIQUES
- Abdominal/pelvic ultrasound
- CT scan
- MRI scan
- Labelled white cell scan
- Mandible/sinus X-ray
- Echocardiography

Repeat history & examination

No diagnosis

or

Diagnosis

Trial of treatment

Definitive treatment

No response ← Response

MOLECULAR DIAGNOSIS
- 16S and 18S rRNA NAAT on any pus sample

Medical Microbiology and Infection at a Glance, Fourth Edition. Stephen H. Gillespie, Kathleen B. Bamford. © 2012 John Wiley & Sons, Ltd.

Pyrexia of unknown origin

Definition

An intermittent or persistent fever of >38.2°C lasting for more than 2 weeks and for which there is no obvious cause.

Aetiology

• Infection accounts for 45–55% (e.g. endocarditis, tuberculosis, hidden abscess).
• Malignancy accounts for 12–20% of cases (e.g. lymphomas or renal cell carcinoma).
• Connective tissue disorders account for 10–15% (e.g. systemic lupus erythematosus or polyarteritis nodosum).
• Other causes to be considered include hypersensitivity to drugs, pulmonary emboli, granulomatous diseases (e.g. sarcoidosis), metabolic conditions (e.g. porphyria) and a factitious cause (i.e. fever induced deliberately by the patient).
• The longer the history of fever, the more likely it is that it arises from a **non-infectious** cause.

Investigation

• A detailed history of the presenting complaint.
• Occupational and social history is important (e.g. vets and farmers are at risk of zoonotic infections, such as brucellosis).
• Travel may suggest exposure to unusual tropical infections.
• Sexual history may indicate possible HIV or related risks.
• Drug history must include those prescribed by a doctor or not, as some drugs and health products can cause fever.
• Patients may overlook vital information at the first interview; further opportunities for discussion are often necessary.
• The patient must be examined carefully for localized bone or joint pain, subtle rashes, lymphadenopathy, abdominal masses, cardiac murmurs and mild meningism.
• The physical examination should be repeated regularly to detect changes, such as development of a new cardiac murmur or enlargement of the liver that becomes palpable only at a later examination.

After the initial history and examination is taken the investigative process can be divided into three phases.

1 Screening investigations, which are performed on all patients (see Figure).
2 Investigations that depend on the results of history and physical examination.
3 Further screening tests and special imaging techniques: ultrasound, CT, MRI, echocardiography and dental X-ray.

The results of the preliminary history and examination are taken together with the results of the primary investigations to plan the tests that will be performed in the second round. These investigations are chosen based on syndrome groups (e.g. fever, eosinophilia and tropical travel). If a diagnosis is not made, further imaging techniques may be required.

Management

A diagnosis should be made before any antimicrobial therapy is commenced. In some infections, notably tuberculosis, which may be suspected but not proven, a trial of therapy may be considered. If this produces clinical improvement, a full course of chemotherapy may be initiated.

Septicaemia

Sources

• **Skin sepsis:** *Streptococcus pyogenes* or *Staphylococcus aureus* bacteraemia may complicate skin infections.
• **Normal flora:** The organisms in the normal flora may become invasive (e.g. dental abscess, cholecystitis, appendicitis or diverticulitis).
• **Following surgery:** A wide range of organisms may be involved. This problem is particularly important in surgery that involves prosthetic devices (orthopaedic, cardiovascular, neurosurgical).
• **Urinary tract:** Enterobacteriaciae or enterococci (see Chapter 51).
• **Pneumonia:** *Streptococcus pneumoniae* (see Chapter 50).
• **Meningitis:** *Neisseria meningitidis* or *S. pneumoniae.*

Clinical features

• Patients are usually severely ill with fever, shock and, sometimes, depressed consciousness.
• If fever is absent, more common in children and elderly people, or shock has not yet developed, patients may present with confusion, drowsiness or being generally unwell.
• It is impossible to distinguish Gram-positive from Gram-negative shock.
• Some infections have characteristic signs (e.g. the purpura of *N. meningitidis*).

Diagnosis and treatment

With the exception of suspected *N. meningitidis* infection, at least two specimens of peripheral blood should be taken for culture before therapy is commenced. Other investigations are directed towards finding the source of the sepsis, such as culture of urine, sputum, skin swabs and/or CSF. Chest and abdominal X-rays and abdominal ultrasound should be performed.

Parenteral empirical antibiotics should be chosen based on the source of infection and the likely infecting organism.

Puerperal fever

This severe, usually bacteraemic, infection is caused by entry of pathogens through the placental bed or the cervix within 7 days of delivery. The organisms involved are either primary pathogens introduced by medical manipulation (e.g. *S. pyogenes*) or elements of the normal flora (e.g. *Bacteroides* and coliforms) associated with retained products. Fever, back pain, offensive lochia or shock may be present and the infection may be complicated by disseminated intravascular coagulation (DIC). Fever in the early puerperium should be investigated with blood and urine cultures and endocervical swabs. Empirical treatment with parenteral antibiotics, such as a third-generation cephalosporin and metronidazole or a combination for example of piperacillin plus tazobactam, should begin without delay. Retained products of conception should be removed. Intensive care support may be required.

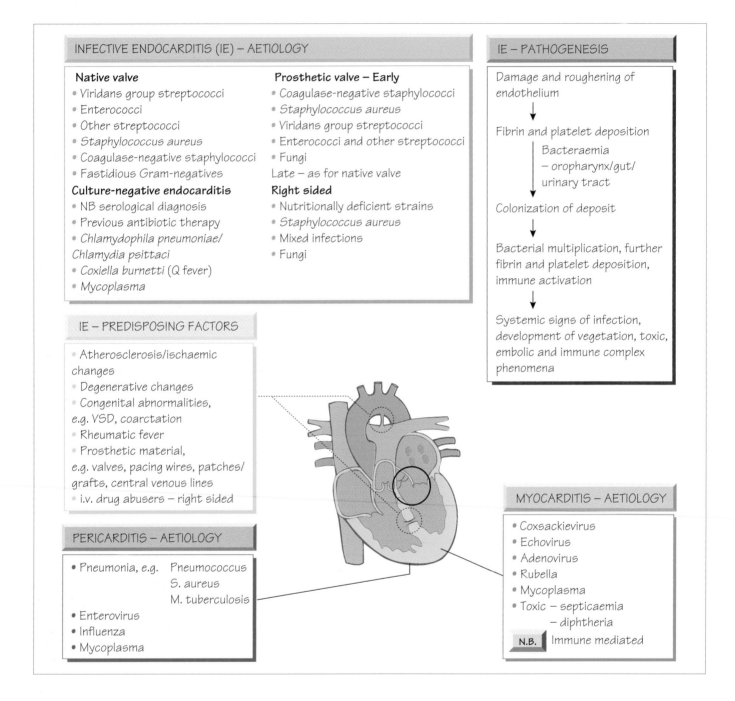

INFECTIVE ENDOCARDITIS (IE) – AETIOLOGY

Native valve
- Viridans group streptococci
- Enterococci
- Other streptococci
- Staphylococcus aureus
- Coagulase-negative staphylococci
- Fastidious Gram-negatives

Culture-negative endocarditis
- NB serological diagnosis
- Previous antibiotic therapy
- Chlamydophila pneumoniae/ Chlamydia psittaci
- Coxiella burnetti (Q fever)
- Mycoplasma

Prosthetic valve – Early
- Coagulase-negative staphylococci
- Staphylococcus aureus
- Viridans group streptococci
- Enterococci and other streptococci
- Fungi
Late – as for native valve

Right sided
- Nutritionally deficient strains
- Staphylococcus aureus
- Mixed infections
- Fungi

IE – PATHOGENESIS

Damage and roughening of endothelium
↓
Fibrin and platelet deposition
　　Bacteraemia
　　– oropharynx/gut/ urinary tract
↓
Colonization of deposit
↓
Bacterial multiplication, further fibrin and platelet deposition, immune activation
↓
Systemic signs of infection, development of vegetation, toxic, embolic and immune complex phenomena

IE – PREDISPOSING FACTORS

- Atherosclerosis/ischaemic changes
- Degenerative changes
- Congenital abnormalities, e.g. VSD, coarctation
- Rheumatic fever
- Prosthetic material, e.g. valves, pacing wires, patches/ grafts, central venous lines
- i.v. drug abusers – right sided

PERICARDITIS – AETIOLOGY

- Pneumonia, e.g.　Pneumococcus
　　　　　　　　　S. aureus
　　　　　　　　　M. tuberculosis
- Enterovirus
- Influenza
- Mycoplasma

MYOCARDITIS – AETIOLOGY

- Coxsackievirus
- Echovirus
- Adenovirus
- Rubella
- Mycoplasma
- Toxic – septicaemia
　　　　 – diphtheria

N.B. Immune mediated

Medical Microbiology and Infection at a Glance, Fourth Edition. Stephen H. Gillespie, Kathleen B. Bamford. © 2012 John Wiley & Sons, Ltd.
Published 2012 by John Wiley & Sons, Ltd.

Endocarditis

Heart valves may become infected during transient bacteraemia. Congenitally abnormal or damaged valves are at greatest risk. Bacteria may originate from the mouth, urinary tract, intravenous drug misuse or colonized intravascular lines.

Clinical features

- Malaise, fever and variable heart murmurs.
- Arthralgia.
- Classical stigmata (e.g. splinter haemorrhages, Osler's nodes, microhaematuria, retinal infarcts, finger clubbing, *café-au-lait* skin, and Janeway lesions) are only seen in long-standing infection.
- In the later stages, septic emboli may cause a stroke.
- With aggressive organisms (e.g. *Staphylococcus aureus*) disease progresses rapidly and valves may rupture.

Diagnosis

A diagnosis is made if there are two major Duke criteria present or one major and three minor criteria.

The major Duke criteria include: positive blood culture with characteristic organisms (e.g. viridans streptococci); persistently positive blood cultures with any organism; evidence of endocardial involvement demonstrated by echocardiogram; and new valvular regurgitation. Minor criteria include: predisposition; fever >38 °C; immunological signs (e.g. septic pulmonary infarcts); and echocardiographic or microbiological evidence insufficient to be a major criterion.

Complications

- Local progression may lead to aortic root abscess.
- Valve destruction may lead to cardiac decompensation.
- Cerebral or limb infarction may follow septic embolus.
- Nephritis secondary to immune complex deposition can progress rapidly if sepsis is uncontrolled or if antibiotics with renal toxicity are given without care (e.g. aminoglycosides).

Investigation

Echocardiography, either transthoracic or transoesophageal (more sensitive), will demonstrate vegetations on the valves; a plain chest X-ray may show evidence of cardiac failure. At least three sets of blood cultures should be taken, an hour apart, while fever is present. Antibiotic therapy should await the results of blood culture if possible. Serum should be tested for antibodies to *Coxiella*, *Bartonella* and *Chlamydia psittaci*.

Management

Ideally, antibiotics should not be commenced until the identity and sensitivities of the infecting organism are known; the prognosis of empirically treated, culture-negative endocarditis is poorer than that when the infecting organism is identified. Careful microbiological monitoring of the markers of inflammation (e.g. CRP) is associated with an improved outcome.

Therapy is based on the minimum inhibitory concentration (MIC) and the minimum bactericidal concentration (MBC) of the antibiotics. If gentamicin is used concentrations must be monitored closely. Therapy is continued for 2–6 weeks depending on the organism and its susceptibility. Typical regimens include: benzylpenicillin and gentamicin for viridans streptococci; flucloxacillin and gentamicin for staphylococci; vancomycin and gentamicin for penicillin-allergic patients. National evidence-based guidelines for management should be followed.

Surgical management may be required to deal with the haemodynamic consequences of endocarditis (e.g. valve rupture).

Prevention

Antibiotic prophylaxis is given to patients with damaged valves when they undergo procedures that give rise to significant bacteraemia, such as dental work or urogenital surgery. For procedures that require an anaesthetic, prophylaxis is given at induction followed by subsequent oral doses. There are alternative regimens for penicillin allergy and prosthetic valves laid down by national guidelines.

Myocarditis

Most myocarditis is caused by viral infection, with enteroviruses being the commonest cause; however, it may complicate systemic viral infections, follow bacteraemia or form part of brucellosis, rickettsial or chronic Chagas infection.

- Presentation is with influenza-like symptoms associated with fatigue, exertional dyspnoea, palpitations and precordial pain.
- Tachycardia, dysrhythmia or cardiac failure may be present.
- The electrocardiogram (ECG) may show T-wave inversion, prolongation of the PR or QRS interval, extrasystoles or heart block.
- Other features include cardiomegaly on chest X-ray and elevated cardiac enzymes.
- The diagnosis is suggested by a temporal relationship between the viral symptoms and cardiological abnormalities.
- Viruses may be detected in faecal, nasopharyngeal and throat specimens (see Chapter 32) by nucleic acid amplification tests (NAATs).
- Treatment is supportive.

Pericarditis

Pericarditis is most often secondary to a non-infectious condition, such as myocardial infarction. It may complicate bacteraemia or follow spread of pus from an empyema (*Streptococcus pneumoniae*), or liver abscess (enterococci, *Entamoeba histolytica*). Tuberculosis can cause sub-acute pericarditis. Viral pericarditis is a self-limiting condition featuring fever, 'flu-like symptoms and sharp chest pain. Enteroviruses, especially coxsackie and influenza viruses, are most commonly implicated. The chest pain may vary with posture, swallowing or heartbeat. A pericardial rub may be heard. Cardiographic evidence of pericarditis may be demonstrated.

Patients with suppurative pericarditis present with fever, neutrophilia and signs of the underlying source of infection. Chest pain is severe and a fall in blood pressure may indicate developing tamponade. The ECG shows upward-curving elevated ST segments. Echocardiography will show pericardial thickening or effusion. Infection can be complicated by fibrosis and constrictive pericarditis, leading to congestive cardiac failure. Treatment is directed against the likely causative organism.

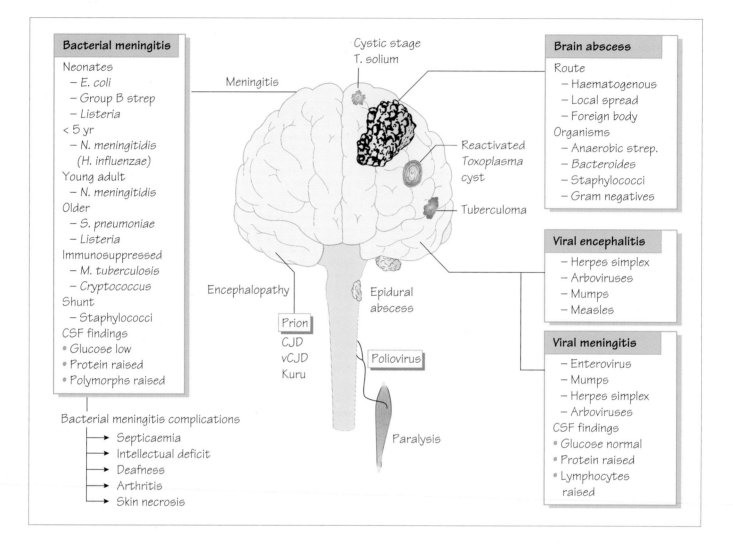

Bacterial meningitis

Neonates
- E. coli
- Group B strep
- Listeria

< 5 yr
- N. meningitidis
  (H. influenzae)

Young adult
- N. meningitidis

Older
- S. pneumoniae
- Listeria

Immunosuppressed
- M. tuberculosis
- Cryptococcus

Shunt
- Staphylococci

CSF findings
- Glucose low
- Protein raised
- Polymorphs raised

Bacterial meningitis complications
→ Septicaemia
→ Intellectual deficit
→ Deafness
→ Arthritis
→ Skin necrosis

Meningitis

Cystic stage
T. solium

Reactivated
Toxoplasma
cyst

Tuberculoma

Encephalopathy

Prion
CJD
vCJD
Kuru

Epidural
abscess

Poliovirus

Paralysis

Brain abscess

Route
- Haematogenous
- Local spread
- Foreign body

Organisms
- Anaerobic strep.
- Bacteroides
- Staphylococci
- Gram negatives

Viral encephalitis
- Herpes simplex
- Arboviruses
- Mumps
- Measles

Viral meningitis
- Enterovirus
- Mumps
- Herpes simplex
- Arboviruses

CSF findings
- Glucose normal
- Protein raised
- Lymphocytes raised

Bacterial meningitis

Clinical features
- Fever, headache, photophobia and neck stiffness.
- Occasionally vomiting and diarrhoea may predominate.
- Levels of consciousness fall progressively.
- Not all signs and symptoms may be present, especially in neonates and elderly people where presentation may be atypical.
- A bulging fontanelle in a neonate indicates raised intracranial pressure.
- Ear or sinus infections may suggest a pneumococcal cause.

Complications
- With appropriate treatment, mortality from *Haemophilus influenzae* and *Neisseria meningitidis* should be below 5%. However, *N. meningitidis* septicaemia can be rapidly fatal if the diagnosis is delayed.

- *H. influenzae* – recurrence of fever, hydrocephalus, convulsions and deafness.
- *N. meningitidis* – deafness, intellectual deficit, skin necrosis and reactive arthritis.
- *Streptococcus pneumoniae* has the highest mortality (>20%) and deafness, cranial nerve palsies and hydrocephalus are complications.

Diagnosis
- CSF should be obtained if possible but treatment of suspected meningitis should not be delayed.
- Total and differential white cell count, Gram stain, Ziehl–Neelsen, India ink, nucleic acid amplification test (NAAT) and antigen detection methods should be used.
- Blood should be taken for culture, rapid antigen detection and glucose determination. Meningococcal disease can be diagnosed by NAAT on whole blood.

Medical Microbiology and Infection at a Glance, Fourth Edition. Stephen H. Gillespie, Kathleen B. Bamford. © 2012 John Wiley & Sons, Ltd.
Published 2012 by John Wiley & Sons, Ltd.

Management

- Neonatal meningitis (likely *Escherichia coli*, group B *Streptococcus* and *Listeria*): cefotaxime and an aminoglycoside, plus ampicillin if *Listeria* is suspected.
- Adult: ceftriaxone 2 g per day.
- If penicillin-resistant *S. pneumoniae* is suspected: vancomycin can be added.
- Cryptococcal meningitis: amphotericin and 5-flucytosine.
- Tuberculous meningitis: rifampicin, pyrazinamide, isoniazid and ethambutol (see Chapter 18).
- Shunt-related meningitis: treatment according to the identity and susceptibilities of the organisms.

Prevention

- Protein-conjugated polysaccharide vaccines are available for *H. influenzae* and *N. meningitidis* serogroups A and C, but not serogroup B, which causes most cases in the UK.
- Family (close) contacts of meningococcal and *Haemophilus* meningitis patients require antimicrobial prophylaxis (ciprofloxacin or rifampicin).

Brain abscess

Brain abscesses arise from parameningeal suppuration, foreign bodies or haematogenous spread from distant sepsis. Infection is polymicrobial with anaerobic cocci, *Prevotella* spp., staphylococci, streptococci (*S. anginosus*/*S. constellatus* group) and Enterobacteriaceae.

Clinical features

- Headache, fever and reduced consciousness.
- Focal neurological signs depend on the location of the abscess.
- Signs of raised intracranial pressure may develop followed by seizures.

Diagnosis and treatment

Lesions are localized by CT scanning. A lumbar puncture is contraindicated because of the risk of cerebellar herniation. Drainage should be performed if feasible and pus sent for culture, NAAT and sensitivity testing.

A regimen of cefotaxime, metronidazole and penicillin or a combination of benzylpenicillin, chloramphenicol and metronidazole may be used.

Viral meningitis

Meningitis and encephalitis may arise from infection with enteroviruses, mumps, herpes simplex, arboviruses, influenza and, rarely, rubella or Epstein–Barr virus (EBV). Viral meningitis can be part of the natural history of polio infection (see Chapter 36). Patients present with headache, photophobia, fever and neck stiffness. The CSF shows an increase in lymphocytes, a mildly raised protein level and normal glucose level. Throat swabs, CSF and stool specimens should be sent for NAAT and serological testing. Management is symptomatic as most patients recover without residual deficit within a few days.

Viral encephalitis

Viral encephalitis is caused by a variety of viruses including herpesvirus and arbovirus (see Chapters 31 and 38). Patients are febrile with headache, neck stiffness and impaired consciousness. Focal neurological signs may develop; convulsions are common. Virus may be detected in CSF, stool and throat specimens by NAAT and serological techniques. Aciclovir is used for treatment of herpetic encephalitis and has led to reductions in both mortality and the number of patients left with severe residual disability.

Postinfectious encephalitis

Some viruses are associated with encephalitis after a systemic infection has resolved, known as postinfectious encephalitis (e.g. measles, varicella zoster, rubella, EBV, mumps and influenza). Clinically similar to viral meningitis, this condition is thought to be mediated by an autoimmune reaction.

Spongiform encephalopathies

The prion protein is a protease-resistant form of a protein that is a normal constituent of the brain. When ingested, the prion protein induces a conformational change in the host brain protein, which leads to spongiform degeneration in the brain. There is an extended incubation period of more than 5 years.

Kuru was a disease seen in Papua New Guinea among people who ate human tissue, including brain. Variant Creutzfeldt–Jakob disease (vCJD) arose after animal brain protein was fed to cattle, the disease being transmitted to humans by ingestion of contaminated beef products. The animal disease, bovine spongiform encephalopathy (BSE), has been eliminated in countries with effective control measures. The size and scale of any human epidemic is uncertain but likely to be small.

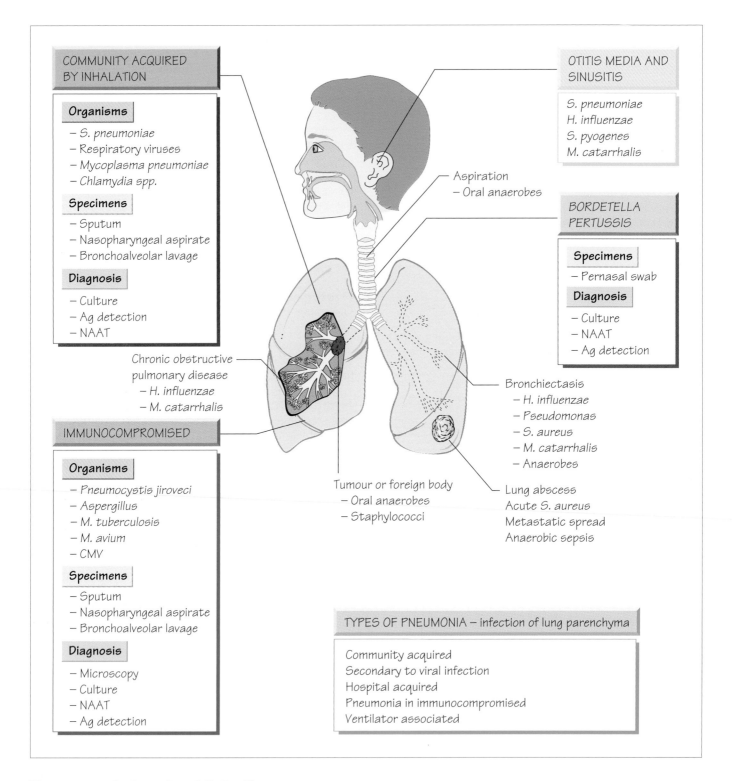

COMMUNITY ACQUIRED BY INHALATION

Organisms
- S. pneumoniae
- Respiratory viruses
- Mycoplasma pneumoniae
- Chlamydia spp.

Specimens
- Sputum
- Nasopharyngeal aspirate
- Bronchoalveolar lavage

Diagnosis
- Culture
- Ag detection
- NAAT

Chronic obstructive pulmonary disease
- H. influenzae
- M. catarrhalis

IMMUNOCOMPROMISED

Organisms
- Pneumocystis jiroveci
- Aspergillus
- M. tuberculosis
- M. avium
- CMV

Specimens
- Sputum
- Nasopharyngeal aspirate
- Bronchoalveolar lavage

Diagnosis
- Microscopy
- Culture
- NAAT
- Ag detection

OTITIS MEDIA AND SINUSITIS

S. pneumoniae
H. influenzae
S. pyogenes
M. catarrhalis

Aspiration
- Oral anaerobes

BORDETELLA PERTUSSIS

Specimens
- Pernasal swab

Diagnosis
- Culture
- NAAT
- Ag detection

Bronchiectasis
- H. influenzae
- Pseudomonas
- S. aureus
- M. catarrhalis
- Anaerobes

Tumour or foreign body
- Oral anaerobes
- Staphylococci

Lung abscess
Acute S. aureus
Metastatic spread
Anaerobic sepsis

TYPES OF PNEUMONIA – infection of lung parenchyma

Community acquired
Secondary to viral infection
Hospital acquired
Pneumonia in immunocompromised
Ventilator associated

Upper respiratory tract infections
Pharyngitis
This is a common condition seen in community practice. Patients have fever and a painful, infected throat that may have visible pus or exudate. Regional lymph nodes may be painful and enlarged.

Bacterial and fungal causes include *Streptococcus pyogenes*, *Neisseria gonorrhoeae* and *Candida*. *Corynebacterium diphtheriae* infection should be considered if there is an appropriate travel or vaccination history (see Chapter 11). Streptococcal infection may be complicated by peritonsillar abscess (quinsy), bacteraemia, rheumatic fever or nephritis.

Medical Microbiology and Infection at a Glance, Fourth Edition. Stephen H. Gillespie, Kathleen B. Bamford. © 2012 John Wiley & Sons, Ltd.
 Published 2012 by John Wiley & Sons, Ltd.

Management
- Most infections in adults are viral so symptomatic treatment is usually adequate.
- Penicillin V or a macrolide is given if bacterial infection is suspected or proven by near-patient testing.
- Ampicillin should be avoided as it may provoke a rash with Epstein–Barr virus (EBV) infection.
- Tonsillectomy or adenoidectomy may reduce the frequency of episodes of pharyngitis or otitis media in patients with quinsy or recurrent otitis media.

Otitis media and sinusitis
- Infection occurs when the Eustachian tube or sinuses are occluded by inflammation.
- Children under 7 years are especially prone because the Eustachian tube is short, narrow and nearly horizontal.
- *Streptococcus pneumoniae, S. pyogenes, Haemophilus influenzae, Moraxella catarrhalis* and the more recently recognized *Alloiococcus otitidis* are the commonest causative organisms.
- Presentation is with local pain and fever.
- With sinusitis, the pain may be worse with head movement and in the evening.
- Ear infection may be complicated by perforation, recurrent or chronic infection or the development of 'glue ear' (sterile mucus within the middle ear).
- Rarely, acute meningitis or mastoiditis can complicate severe infection.
- Diagnosis is clinical: an auroscope may show retrotympanic fluid levels, an inflamed tympanic membrane or a purulent discharge associated with perforation.
- Treatment depends on reducing the mucosal swelling, promoting drainage of the fluid and encouraging the recirculation of air.
- Appropriate antibiotic therapy is used in more severe cases.

Acute epiglottitis
This infectious swelling of the epiglottis may threaten the airway. *H. influenzae* type b was a common cause until vaccination became available. Infection with *S. pyogenes* causes some cases, usually in adults.
- Presentation is with sore throat and high fever.
- Stridor and drooling are usually present.
- Throat examination should be avoided because it may precipitate acute respiratory obstruction.
- Treatment is with parenteral third-generation cephalosporins.
- Emergency tracheostomy may become necessary.

Lower respiratory tract infections
Epidemiology
Infections of the lower respiratory tract are an important cause of morbidity and mortality worldwide and a major cause of death in children under 5 years.

Predisposing factors include:
- smoking;
- chronic obstructive pulmonary disease;
- diabetes mellitus;
- immunosuppressive therapy;
- HIV.

The bacterial causes are illustrated and many viruses cause primary viral pneumonia (e.g. influenza and SARS coronavirus).

Others cause damage to the lower respiratory tract, permitting secondary bacterial pneumonia.

Clinical features
- Fever and a cough.
- Purulent sputum production that may be blood stained.
- Some pathogens (e.g. *Mycoplasma*) rarely cause productive cough.
- Pleural inflammation causes sharp chest pain that is worse on inspiration.
- Signs of systemic infection, such as myalgia, malaise and weakness, may be present.
- In elderly people, mental confusion is common even when specific symptoms and signs are slight.

Complications
- Pleural and pericardial spread.
- Septicaemia.
- Meningitis.
- *Staphylococcus aureus* infection can be complicated by lung cavitation and bronchiectasis.

Diagnosis
Diagnostic samples include:
- expectorated sputum;
- induced sputum;
- bronchoalveolar lavage (especially for suspected tuberculosis or for immunocompromised patients).

Rapid diagnostic techniques may be possible, such as urinary antigen detection for *S. pneumoniae* and *Legionella pneumophila*)

Multiplex nucleic acid amplification tests (NAATs) are available for all respiratory pathogens and can be performed quickly enough to inform treatment choice.

Management and prevention
- Appropriate antibiotic therapy should be started as soon as possible.
- Severe, community-acquired pneumonia requires hospitalization with intravenous antibiotics (e.g. a third-generation cephalosporin and macrolide).
- Milder infections can be treated with oral therapy, often with amoxicillin and/or a macrolide, although quinolones such as moxifloxacin are also used. As β-lactam resistance is common in *H. influenzae*, patients with chronic obstructive pulmonary disease should be treated with an appropriate agent (e.g. co-amoxiclav or trimethoprim).
- Treatment of hospital-acquired pneumonia may require agents that are active against Enterobacteriaceae and *Pseudomonas* (e.g. ciprofloxacin or ceftazidime).
- Supportive therapy may include bed-rest, oxygen, rehydration, physiotherapy and ventilation if needed.

Infective exacerbations of cystic fibrosis are often initially with *H. influenzae*; later infections with *Pseudomonas* and *Burkholderia cepacia* require specialist management with detailed culture and susceptibility testing that allows the optimization of antimicrobial therapy. This should be coupled with intensive postural drainage and physiotherapy.

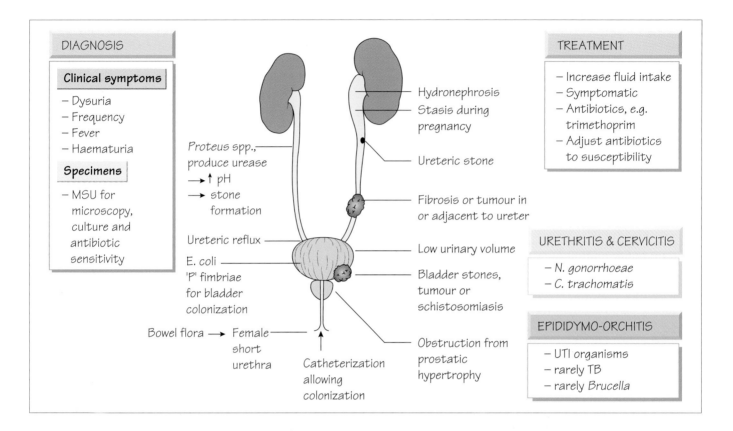

DIAGNOSIS

Clinical symptoms
- Dysuria
- Frequency
- Fever
- Haematuria

Specimens
- MSU for microscopy, culture and antibiotic sensitivity

Proteus spp., produce urease
→ ↑ pH
→ stone formation

Ureteric reflux
E. coli 'P' fimbriae for bladder colonization

Bowel flora → Female short urethra

Catheterization allowing colonization

Hydronephrosis
Stasis during pregnancy

Ureteric stone

Fibrosis or tumour in or adjacent to ureter

Low urinary volume

Bladder stones, tumour or schistosomiasis

Obstruction from prostatic hypertrophy

TREATMENT
- Increase fluid intake
- Symptomatic
- Antibiotics, e.g. trimethoprim
- Adjust antibiotics to susceptibility

URETHRITIS & CERVICITIS
- N. gonorrhoeae
- C. trachomatis

EPIDIDYMO-ORCHITIS
- UTI organisms
- rarely TB
- rarely Brucella

Urinary tract infection

Anatomical considerations

Only the lower part of the urethra is usually colonized by bacteria; the flushing action of urinary flow prevents ascending infection. The shorter female urethra makes urinary infection more common.

Epidemiology and pathogenesis

Dehydration, obstruction, the disturbance of smooth urinary flow or the presence of a foreign body such as a stone or urinary catheter predisposes to urinary infection. Trauma during sexual intercourse may precipitate infection in women. Infection in children, especially in boys, is often associated with congenital abnormalities, such as ureteric reflux or urethral valves.

Clinical features

- **Lower urinary tract infections:** urinary frequency, dysuria and suprapubic discomfort; fever may be absent.
- **Pyelonephritis:** fever, loin pain, renal angle tenderness and signs of septicaemia.
- **Infection in children, elderly people and antenatal patients:** may be clinically silent.
- Recurrent infections can result in scarring and renal failure.

Laboratory diagnosis '

- 'Dip-stick' test for leukocyte esterase and nitrite can identify patients with infection and the need for treatment without culture.
- Culture using a midstream urine (MSU) specimen to reduce the risk of contamination.
- $>10^5$ colony-forming units/mL of a single organism indicate infection, whereas $<10^5$ organisms/mL or a mixed growth suggests contamination.
- Chronically catheterized patients usually have 'significant' numbers of organisms and multiple pathogens and may not have active infection.
- All isolates are potentially significant from a suprapubic aspirate from an infant with suspected infection.
- Susceptibility tests should be performed on all significant isolates.

Treatment and prevention

- Empirical therapy is based on the known susceptibilities of urinary pathogens.
- Most community-acquired infections respond to oral antibiotics (e.g. cefalexin, amoxicillin, trimethoprim or nitrofurantoin).
- If septicaemia is present, ciprofloxacin or cefotaxime should be used.

Medical Microbiology and Infection at a Glance, Fourth Edition. Stephen H. Gillespie, Kathleen B. Bamford. © 2012 John Wiley & Sons, Ltd.

- Recurrent urinary infection may require nocturnal prophylaxis (e.g. low-dose trimethoprim, nitrofurantoin or naladixic acid), together with advice on ensuring an adequate urine flow is achieved.
- Children with recurrent infection should be investigated for anatomical abnormalities.
- Significant bacteriuria in pregnant women should be treated, even if asymptomatic.
- Anatomical obstructions to urine flow should be removed if possible.

Genital infection

Genital infection presents in many ways (see Table). It may be followed by pelvic inflammatory disease, infertility, prostatitis, arthritis or bacteraemia. Other sites may be involved, for example the throat and rectum in gonococcal infection.

Genitourinary infection syndromes and causative organisms.	
Syndrome	Organisms
Genital ulcers	Herpes simplex
	Chlamydia trachomatis types L1–4
	Haemophilus ducreyi (see Chapter 22)
	Treponema pallidum (see Chapter 28)
	Calymmatobacterium donovani
Urethral discharge	*Neisseria gonorrhoeae*
	C. trachomatis
Pelvic inflammatory disease	*N. gonorrhoeae*
	C. trachomatis
	Mixed anaerobic infection
Vaginal discharge	*Candida albicans*
	Trichomonas vaginalis
	Mobiluncus spp. and others in non-specific vaginitis

Diagnosis

Urethral and cervical swabs should be taken for both bacterial and viral diagnosis. *N. gonorrhoeae*, *Chlamydia* and herpes simplex are optimally detected by nucleic acid amplification tests (NAATs; see Chapters 27 and 31). Samples positive for *N. gonorrhoeae* can be cultured for susceptibility testing. Syphilis is diagnosed with enzyme immunoassay (EIA) together with traditional treponemal tests (see Chapter 28). Direct microscopy may show evidence of *Candida* or *Trichomonas*.

Treatment

Patients must be treated before a laboratory diagnosis, so treatment is guided by a 'syndromic approach' where therapy is based on the agents that are likely to treat at least 95% of organisms in a community. For example, patients with uncomplicated urethritis can be treated with a single dose of a suitable cephalosporin or fluoroquinolone followed by a 1-week course of either doxycycline or single-dose azithromycin. Syphilis is treated with penicillin (see Chapter 28).

Prevention

- Risk avoidance (e.g. monogamous relationships).
- Risk reduction (e.g. barrier contraceptive methods).
- Tracing of sexual contacts to treat asymptomatic disease.
- Antigen variability in *N. gonorrhoeae* means that there is no effective vaccine for gonorrhoea.

Trichomonas vaginalis

- Causes an itchy vaginal infection with an offensive discharge.
- Treatment of sexual contacts may be necessary to prevent recurrent infection.

Non-specific vaginosis

- Caused by disruption to the normal vaginal flora.
- Results in an offensive discharge with a characteristic fishy smell when alkalinized.
- Diagnosis is based on clinical findings and near-patient tests (e.g. 'clue cells', which are epithelial cells heavily coated with bacteria, and a positive amine test) and defined syndrome scoring schemes.
- Clindamycin preparations, oral metronidazole, and oral and intravaginal tablets of lactobacillus are effective treatments.

Epididymo-orchitis

Infection of the epididymis may arise (i) from a urinary tract infection, (ii) as part of a genital infection or (iii) as a primary systemic infection, such as brucellosis or tuberculosis. Patients present with a painful, acutely inflamed epididymis and testis, which must be differentiated from testicular torsion. Diagnosis is made clinically and confirmed by the result of urinary or blood cultures and tests for sexually transmitted infections.

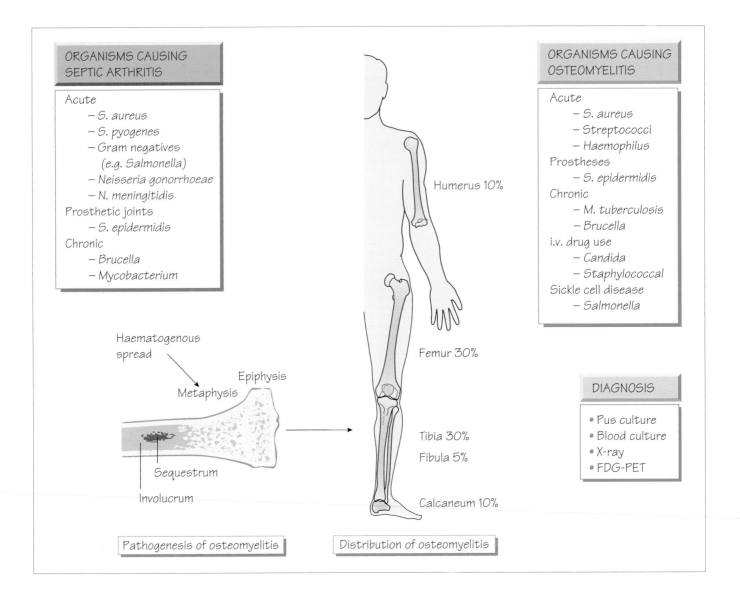

ORGANISMS CAUSING SEPTIC ARTHRITIS

Acute
— S. aureus
— S. pyogenes
— Gram negatives
  (e.g. Salmonella)
— Neisseria gonorrhoeae
— N. meningitidis
Prosthetic joints
— S. epidermidis
Chronic
— Brucella
— Mycobacterium

ORGANISMS CAUSING OSTEOMYELITIS

Acute
— S. aureus
— Streptococci
— Haemophilus
Prostheses
— S. epidermidis
Chronic
— M. tuberculosis
— Brucella
i.v. drug use
— Candida
— Staphylococcal
Sickle cell disease
— Salmonella

Haematogenous spread

Epiphysis
Metaphysis
Sequestrum
Involucrum

Humerus 10%

Femur 30%

Tibia 30%
Fibula 5%

Calcaneum 10%

DIAGNOSIS
• Pus culture
• Blood culture
• X-ray
• FDG-PET

Pathogenesis of osteomyelitis

Distribution of osteomyelitis

Osteomyelitis

• Bone can be infected by haematogenous spread, by direct extension from an infected joint, or following trauma, surgery or instrumentation.
• The formation of pus precipitates ischaemia and necrosis; the central area of dead bone is known as the **sequestrum**.
• New bone (the **involucrum**) may form around the infection site.
• In children, the metaphysis of the long bones (femur, tibia and humerus) are most often involved. Infection of the spine is common in adults.
• *Staphylococcus aureus* accounts for 90% of infections; rarer causes include *Streptococcus pyogenes* (4%), *Haemophilus influenzae* (4%), *Escherichia coli*, *Salmonella* spp., *Mycobacterium tuberculosis* and *Brucella*.
• Patients with sickle cell disease are especially prone to *Salmonella* infection.

Clinical features

• Presentation is with fever and pain. In the young, pain may be poorly localized but they may stop moving the affected limb (**pseudoparalysis**).
• Later, soft-tissue swelling may occur and can be followed by sinus formation.
• Pathological fractures may develop.
• Delayed treatment increases the risk of development of chronic osteomyelitis.
• Infection may develop around foreign bodies (e.g. surgical nails or debris from trauma).

Diagnosis

• Diagnosis is clinical.
• Radiological changes develop late in the course of infection and many not be seen.

Medical Microbiology and Infection at a Glance, Fourth Edition. Stephen H. Gillespie, Kathleen B. Bamford. © 2012 John Wiley & Sons, Ltd.

- Isotope scans indicate sites of inflammation.
- Fluorodeoxyglucose positron emission tomography (FDG-PET) is more sensitive.
- Blood cultures and pus from bone obtained via needle or open biopsy allow culture for pathogen identification and susceptibility testing.

Treatment
- Drainage and excision of the sequestrum is supplemented by antibiotic therapy (e.g. flucloxacillin and fusidic acid pending culture results).
- If *Salmonella* is isolated or suspected, ciprofloxacin may be used.
- Treatment lasts for 6 weeks or until there is evidence that inflammation has disappeared and the bone has healed.

Chronic osteomyelitis
- Follows inadequately treated acute infection, or may be secondary to surgery or a fracture. Infection of prosthetic materials with organisms with reduced virulence, such as coagulase-negative staphylococci (CoNS) is increasingly common.
- *S. aureus* is implicated in 50% of cases; the remainder are associated with Gram-negative pathogens (e.g. *Pseudomonas*, *Proteus* and *E. coli*).
- On-going pain, swelling, deformity and/or a chronically discharging sinus may be found.
- Culture-based diagnosis is essential.
- A prolonged course of appropriate antibiotics should accompany appropriate surgery.
- Infected prosthetic devices are usually removed.

Suppurative arthritis
- Follows bacteraemia or injection of the joint.
- 95% of cases are caused by *S. aureus* and *S. pyogenes*. Other causes include Enterobacteriaceae, *Neisseria gonorrhoeae*, *H. influenzae*, *Salmonella* spp., *Brucella* spp., *Borrelia burgdorferi*, *Pasteurella* and *M. tuberculosis*.
- Large joints (e.g. the knee) are most commonly affected.
- Prosthetic joints are at risk of early and late infections (see below).

Clinical features
- Pain, swelling and reduced movement.
- In adults, the onset may be insidious; a history of recent urinary infection or salmonellosis may be reported.
- Cellulitis or specific signs, such as gonococcal rash may be found.
- Septic arthritis must be differentiated from acute rheumatoid arthritis, osteoarthritis, gout, pseudogout or reactive arthritis.
- A diagnostic tap will yield cloudy fluid, and Gram stain and white blood cell count may suggest infection that can be confirmed by culture or nucleic acid amplification test (NAAT).
- Culture for brucellosis should be performed if the history is suggestive.

Intravenous antibiotics that are appropriate to the infecting organisms, either isolated or suspected, should be commenced, and oral therapy is continued for up to 6 weeks. Aspiration and irrigation of the joint may be helpful in severe cases as it reduces inflammatory damage.

Viral arthritis
Some viruses are associated with arthritis, for example parvovirus, rubella, mumps and hepatitis B. Rubella-related arthritis is more common in females and develops a few days after the rash. Several of the alphaviruses cause severe bone and joint symptoms. Reactive arthritis caused by an immune response to the pathogen can follow recovery, for example after meningococcal disease, or *Shigella* or *Chlamydia* infection. The latter can be associated with uveitis and is known as Reiter's syndrome.

Prosthetic joint infections
Prosthetic joints may become infected at the time of operation (early presentation) or as a result of haematogenous spread (later presentation). In early presentation the organisms are from the skin (e.g. *S. epidermidis* and *S. aureus including* MRSA).

Treatment is with intravenous antibiotics, depending on the susceptibility of the infecting organisms. Infection usually results in loss of the prosthesis. Prevention of infection by control measures in both the ward and theatre and antibiotic prophylaxis with an agent active against *S. aureus* (see Chapter 13) is vital.

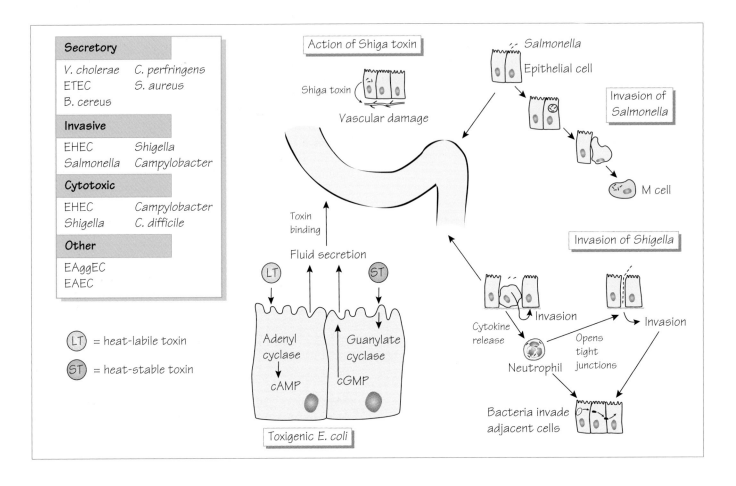

Infectious diarrhoea is common and a very important cause of morbidity and mortality in children under 5 years worldwide. The gut is protected by gastric acid, bile salts, the mucosal immune system and inhibitory substances that are produced by the normal flora.

Epidemiology

Organisms are transmitted by hands and fomites (faecal–oral route), by food or water. The infective dose can be as few as 10 organisms (*Shigella*). Some foods (e.g. milk) or drugs (e.g. H_2 antagonists and proton pump inhibitors) may reduce the protective effects of gastric acid.

Bacteria enter the food chain from infected animals, from poor hygiene during slaughter, and during butchering. Hens that are chronically colonized with *Salmonella* may produce eggs that are contaminated. Improper cooking and storage may allow the multiplication of bacteria (see below). Transmission is favoured by poor sanitation; infection can spread rapidly causing significant mortality. Cholera is capable of spreading worldwide (a pandemic).

Travellers' diarrhoea usually develops within 72 h of arrival in a new country; Latin America, Africa and Asia are the regions with the highest risks. Patients pass two to four watery bowel motions daily; blood and mucus are typically absent. The major organisms implicated are enterotoxigenic *Escherichia coli* (ETEC) and enteroadherent *E. coli* (EAEC; see Chapter 23). Treatment is with fluid replacement and antibiotics, including co-trimoxazole or ciprofloxacin.

Pathogenesis

• Toxin-mediated dysregulation of intestinal cells causing fluid secretion.

• Invasion of the intestinal wall occurs with destruction of the cells (see Figure).

• Secretory diarrhoea produces infrequent large-volume stools as the absorptive capacity of the colon is overwhelmed.

• In dysenteric illness (e.g. *Shigella*), colonic inflammation causes a loss in bowel capacity and frequent, often blood-stained stools.

Medical Microbiology and Infection at a Glance, Fourth Edition. Stephen H. Gillespie, Kathleen B. Bamford. © 2012 John Wiley & Sons, Ltd.

• Enterohaemorrhagic *E. coli* (EHEC) produce Shiga toxin (Stx), which causes bloody diarrhoea and the haemolytic uraemic syndrome (HUS), which is commonest with serotype O157:H7.

Clinical features
• There may be many small stools (typical of large-bowel infection) or infrequent large stools (small-intestinal infection). Stools may be blood stained when there is destruction of the intestinal mucosa, or have a fatty consistency and offensive smell if malabsorption is present.
• Dehydration and electrolyte imbalance may develop rapidly with potentially fatal results, especially in cholera. Crampy abdominal pain may accompany diarrhoea (e.g. *Campylobacter* and *Shigella* infections); this may mimic acute abdominal conditions, such as appendicitis.
• Fever is not always present.
• Septicaemia can occur in salmonellosis, but is rare in other diarrhoeal diseases.
• Secondary lactose intolerance, which is caused by loss of intestinal lactase, may persist for a few weeks.
• Patients with IgA deficiency may have difficulty eradicating *Giardia*; those with T-cell deficiency are prone to *Salmonella* and *Cryptosporidium* (see Chapter 55).

Diagnosis
• Diagnosis is by culture using a range of media specific to different groups of pathogens.
• Multiplex nucleic acid amplification tests (NAATs) are being introduced into routine practice.
• Organisms should be typed using molecular methods for epidemiological purposes (see Chapter 1).
• Toxin may be detected in stool samples (e.g. *Clostridium difficile* toxin).

Management
The management of diarrhoeal disease is based on adequate fluid replacement and correction of electrolyte imbalances. Despite the high outflow found in secretory diarrhoea, fluid absorption still occurs. Oral rehydration solutions that consist of 150–155 mmol/L sodium and 200–220 mmol/L glucose can be life-saving. Intravenous fluid replacement is rarely necessary. Antimotility drugs are of no benefit and may be dangerous, especially in small children. Oral antibiotics, such as tetracycline or ciprofloxacin, which shorten the duration of symptoms, may be of benefit in cholera and other cases of severe fluid diarrhoea. Patients with severe dysentery and salmonellosis should be treated with ciprofloxacin or co-trimoxazole. Renal failure due to HUS following *E. coli* O157 requires specialist management.

Prevention
• Good sanitation is essential in preventing diarrhoeal disease.
• Animal husbandry and slaughter methods should be designed to prevent the introduction of animal intestinal pathogens into the human food chain.
• Food must be cooked to a sufficiently high temperature to kill pathogens and, if not eaten immediately, refrigerated at a low enough temperature to prevent any bacteria multiplying.
• Cooked food should be physically separated from uncooked food to prevent cross-contamination. This is especially true in institutional cooking (e.g. hospitals and restaurants), where many individuals might become infected following a single failure of hygiene.
• Travellers' diarrhoea can be reduced by careful choice of food while travelling.
• Oral, heat-killed and live attenuated cholera vaccines are licensed for use but the protection they provide is short-lived.
• Whole-cell vaccines, purified Vi polysaccharide vaccine and oral Ty21a vaccine are available against typhoid. New Vi conjugate vaccines are being developed.

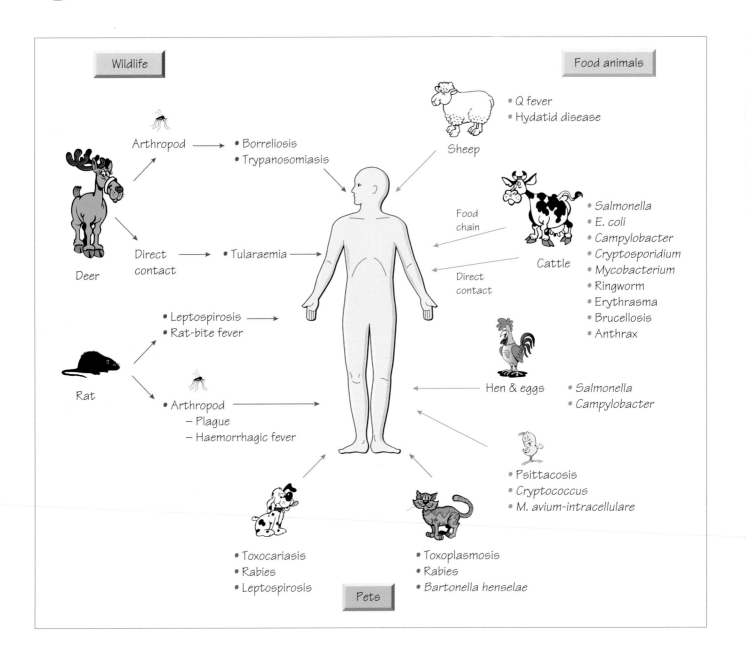

A zoonosis is an infection acquired from an animal source.
• Infection can occur when humans enter the animal environment (e.g. when camping).
• Transmission can be via vectors such as mosquitoes (e.g. Japanese B encephalitis).
• Farming and pets are an important source of zoonotic infection.

Viral zoonoses

More than 100 animal viruses can cause human disease (e.g. Herpes simiae, a monkey pathogen, which causes severe encephalitis, or avian 'flu, both of which are associated with high mortality). Other viral zoonoses are discussed in Chapter 38.

Rat-bite fever

Rat-bite fever is caused by either *Streptobacillus moniliformis* or *Spirillum minus* inoculated by the bite of a rat.
• The incubation period is 3 days to 3 weeks.
• Inflammation occurs at the site of the bite, with associated lymphangitis and regional lymphadenopathy.
• Generalized maculopapular rash, fever, headache and malaise may occur.
• Polyarthritis occurs in 70% of cases.
• Endocarditis is the most serious complication.
• Diagnosis based on bacterial isolation or nucleic acid amplification test (NAAT).
• Treatment is usually with penicillin.

Medical Microbiology and Infection at a Glance, Fourth Edition. Stephen H. Gillespie, Kathleen B. Bamford. © 2012 John Wiley & Sons, Ltd.

Anthrax
See Chapter 16.

Plague
Caused by *Yersinia pestis*, infection is endemic in rodents in remote rural areas. Rarely, epidemics may develop that spread worldwide (e.g. the 'Black Death'). The organism is transmitted both between rats and to humans by the rat flea, *Xenopsylla cheopis*.

The incubation period is short; the disease has an abrupt onset characterized by fevers and toxaemia. The regional lymph glands draining the site of the bite become greatly enlarged (buboes) and septicaemia is accompanied by generalized haemorrhage. Pneumonic plague is a rapidly fatal pneumonitis that can be transmitted by the respiratory route.

Plague is diagnosed clinically in areas where it is endemic. Direct smear of a lymph gland aspirate, blood culture and NAAT are used for diagnosis. Treatment is with tetracycline, chloramphenicol, aminoglycosides or ciprofloxacin. The mortality rate of pneumonic plague is high. There are concerns that it may be used as a bioterrorism weapon.

Borreliosis
Borreliosis is transmitted from rodents or deer by ticks particularly in open forest habitat (e.g. the New Forest) or by lice (see Chapter 28).

Toxoplasmosis
The cat is the definitive host of *Toxoplasma gondii* but the organism infects a wide range of animals, including sheep, cattle and humans. Infection is acquired by ingestion of oocysts from infected cat faeces or from tissue cysts in infected meat (e.g. undercooked beef).

Dermatophytes
Dermatophytes that are natural pathogens of animals can spread to the human population by direct contact (see Chapter 40).

Toxocariasis
Toxocara canis is an ascarid parasite of dogs. The parasite eggs are excreted in the faeces of infected dogs and mature in the soil. Human ingestion occurs when food is contaminated by the soil or when personal hygiene is poor (e.g. poor hand washing). The larval stages hatch in the intestine, invade the host and migrate to the liver and lungs. They are unable to develop into adults but migrate throughout the body causing fever, hepatosplenomegaly, lymphadenopathy and wheeze. If the larva migrates into the eye, sight may be permanently damaged by the local inflammatory response of the retina. The diagnosis is made serologically using a specific enzyme immunoassay (EIA). The disease is usually self-limiting but, if symptoms are severe, treatment with albendazole may be beneficial. Ocular lesions should first be treated with steroids to diminish the inflammatory response; the role of antihelminthic treatment is less certain.

Cat-scratch disease
• Caused by infection with Bartonella henselae.
• There is a 10-day incubation period following a cat scratch or bite.
• A papular lesion may develop at the site.
• Regional lymphadenopathy occurs.
• Symptoms resolve slowly over a period of 2 months, but a more chronic course may ensue.
• Disseminated infection occurs more commonly in immunocompromised individuals.
• The diagnosis is usually made clinically but it can be confirmed serologically by immunofluorescence or EIA. Culture requires a prolonged incubation period; NAATs may also be used for diagnosis.
• Treatment with azithromycin, tetracyclines or rifampicin may be beneficial.

Hydatid disease
Two species of parasite are responsible for human hydatid disease: *Echinococcus granulosus* and *Echinococcus multilocularis*. Dogs are the definitive host of *E. granulosus*, harbouring the tapeworm stage. The eggs, which are passed in the faeces, are ingested by the intermediate hosts (e.g. sheep or rodents) and multiple cysts develop in the liver and lungs. The cycle is completed when dogs eat infected tissues. Humans are accidental hosts. The disease is common in sheep-farming areas.

Echinococcus multilocularis is found in foxes, wolves and dogs; rodents act as the intermediate hosts.

Pathogenesis and clinical features
Cysts act as space-occupying lesions in the liver, lungs, abdominal cavity or central nervous system. The cysts of *E. multilocularis* lack a definite cyst wall and spread in the tissue.

Diagnosis
Cysts may be demonstrated by ultrasound or CT. EIA for both antibody and antigen is available.

Treatment
If possible, hydatid cysts should be surgically removed. Albendazole is given to kill the germinal layer of the cyst and praziquantel to reduce the viability of protoscolices. Puncture, aspiration, injection of chemicals and re-aspiration (PAIR) is seen as an alternative to surgical excision. Cyst rupture can lead to multiple cysts in the abdomen, or to anaphylaxis.

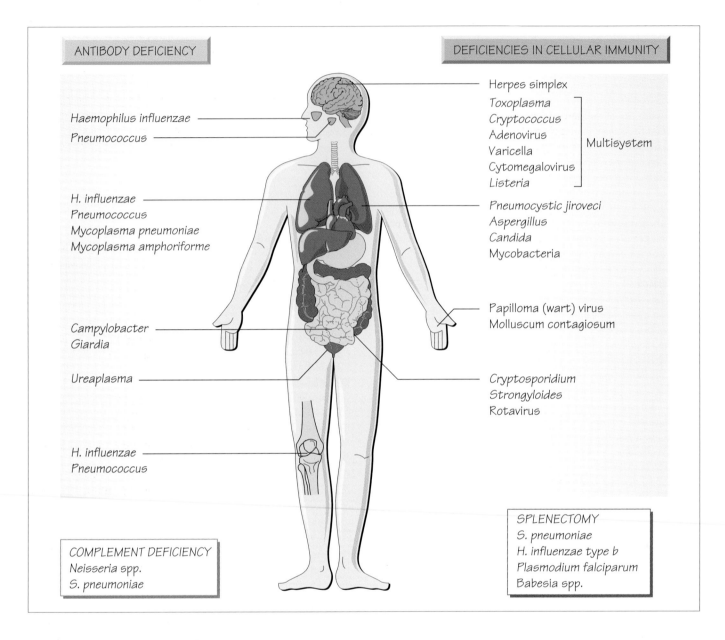

Medical treatment or hereditary deficiency of components of the immune system may allow organisms with reduced virulence to cause infection and normal pathogens to cause severe infection. The origin of immune deficiency is often multifactorial; for example patients undergoing bone marrow transplantation are neutropenic, which reduces resistance to bacterial infection, whilst intravenous cannulation provides a route for *Staphylococcus epidermidis* infection.

Neutropenia

Neutropenia most often arises as a result of acute leukaemia or its treatment; infection risk depends on the duration and severity of the neutropenia. Bacteraemia occurs in between 40 and 70% of neutropenic patients with Enterobacteriaceae and *Pseudomonas*

spp. invading following gut damage by antineoplastic agents or irradiation. Gram-positive organisms (e.g. *S. epidermidis, Streptococcus mitis, Streptococcus oralis, Enterococcus* spp., *Staphylococcus aureus* and *Corynebacterium jeikeium*) are increasingly important causes of sepsis.

With increasing duration of neutropenia, fungal infection may develop due to *Candia albicans, Aspergillus* spp. and, rarely, *Fusarium* spp., *Pseudallescheria boydii* and *Trichosporon beigelii*.

Treatment of fever in neutropenic patients

Empirical therapy includes tazobactam and an aminoglycoside, with vancomycin being added if central-line infection is possible. If fever persists, fungal pathogens are more likely and amphotericin or itraconazole may be added.

Medical Microbiology and Infection at a Glance, Fourth Edition. Stephen H. Gillespie, Kathleen B. Bamford. © 2012 John Wiley & Sons, Ltd.

Prevention of infection

Infection risk can be reduced by:
- source isolation (see Chapter 9);
- filtration of room air to remove fungal spores;
- antifungal prophylaxis (e.g. nystatin, either alone or in combination with oral amphotericin, or fluconazole or itraconazole);
- antibiotic prophylaxis (e.g. fluoroquinolones are used in some centres).

T-cell deficiency

T-cell deficiency is an increasingly common problem, which may follow HIV infection, cancer chemotherapy, corticosteroid therapy or organ transplantation. Congenital T-cell deficiencies are rare but may be purely linked to T-cell function or combined with hypogammaglobulinaemia.

Pathogens

These are mainly the microorganisms that have an intracellular location in the human host, such as:
- *Toxoplasma gondii*, *Strongyloides stercoralis*;
- *Mycobacterium tuberculosis*, *Mycobacterium avium–intracellulare*;
- *Listeria monocytogenes*, *Cryptococcus neoformans*, *Pneumocystis jiroveci*;
- herpes simplex, cytomegalovirus (CMV), varicella zoster virus (VZV) and measles.

Measles infection, especially if complicated by giant cell pneumonia or encephalitis, can be life-threatening.

Diagnosis

Specific infections should be investigated appropriately (see relevant chapters). All patients should have at least two blood cultures taken from different sites.

Hypogammaglobulinaemia

Patients with X-linked agammaglobulinaemia are at increased risk of infection for the first 6 months of life; patients with common variable immunodeficiency are at increased risk throughout life. Functional hypogammaglobulinaemia develops in patients with multiple myeloma.

Patients suffer recurrent respiratory tract infections with *Streptococcus pneumoniae*, the novel *Mycoplasma amphoriforme* and non-capsulate *Haemophilus influenzae*, which lead to bronchiectasis. *Giardia*, *Cryptosporidium* and *Campylobacter* infections may be more persistent. Intravenous immunoglobulin reduces recurrent infection.

Complement deficiency

Hereditary complement deficiencies are rare. Deficiency in the later components of the complement cascade (C7–9) results in an inability to lyse Gram-negative bacteria, and patients are susceptible to recurrent *Neisseria* infection. Deficiency of the alternative complement pathway leads to serious *S. pneumoniae* infections, including meningitis. Acquired complement deficiency occurs in systemic lupus erythematosus.

Mannose-binding lectin

A wide range of bacteria fungi, viruses and protozoa bind to mannose-binding lectin and there are reports that infection with these organisms are commoner or more severe with some deficiency genotypes.

Postsplenectomy infection

The incidence of serious sepsis following splenectomy is about 1% per year; the rate is higher in infants and children. The highest mortality is associated with splenectomy for lymphoma and thalassaemia. Patients with sickle cell disease have functional asplenia. Although the risk of sepsis diminishes with time, it never disappears.

Streptococcus pneumoniae is responsible for approximately two-thirds of infections; others *H. influenzae* and *Escherichia coli*. Malaria may run a fulminant course. Splenectomy predisposes to *Capnocytophaga canimorsus* infection, which usually arises after a dog bite.

Prevention

- Conjugate vaccines against *S. pneumoniae* meningococcus and *H. influenzae* type b.
- Low-dose, oral antibiotic prophylaxis with penicillin V.
- Patient awareness of the urgency of treating respiratory infections with antibiotics, which may be prescribed in advance to avoid delay in initiation of treatment.

Pneumocystis jiroveci

Pneumocystis jiroveci is a fungus causing infection in individuals with severe T-cell dysfunction. Transmitted by the respiratory route, *P. jiroveci* adheres strongly to pneumocytes.

Clinical presentation

- Dyspnoea develops insidiously over days or weeks.
- Patients have an unproductive, dry cough.
- Pleuritic chest pain is uncommon.
- Patients are febrile but chest examination usually normal, although fine basal crackles may be heard.
- Chest X-ray may appear normal initially then develop through reticular shadowing until finally there is diffuse air space consolidation.

Diagnosis

Specimens, obtained by bronchoalveolar lavage or sputum induction may be examined by specific immunofluorescence, or nucleic acid amplification test (NAAT).

Treatment

Treatment is with oral co-trimoxazole in high dosage or intravenous pentamidine. Alternatives include trimethoprim–dapsone, pyrimethamine–clindamycin and atovaquone.

Toxoplasma gondii

Toxoplasma infection persists inside the host cells for very long periods. Falling immunity allows the reactivation of previously dormant infection. A space-occupying lesion may develop in the brain, which may be accompanied by encephalitis.

Toxoplasma encephalitis presents with fever and headaches. Convulsions, coma and focal neurological signs may follow. A CT scan may demonstrate multiple diagnostic focal lesions with ring enhancement. Brain biopsy may yield material for tissue culture or PCR. *Toxoplasma* encephalitis is treated with pyrimethamine–sulfadiazine. Long-term suppressive treatment is required after recovery.

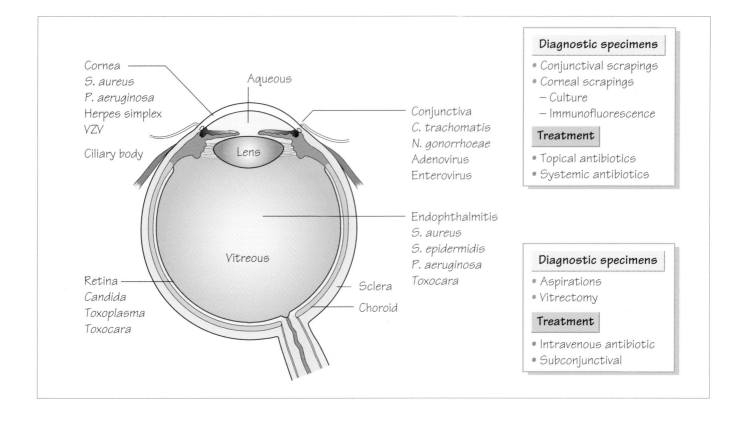

Bacterial conjunctivitis

Bacterial conjunctivitis is common and caused by *Staphylococcus aureus*, *Haemophilus influenzae*, *Streptococcus pneumoniae* or *Moraxella* spp.

Neonatal conjunctivitis is acquired from infection in the mother's genital tract and caused by *Neisseria gonorrhoeae*, *Chlamydia trachomatis*, *Escherichia coli*, *S. aureus* and *H. influenzae*. In hospital if ocular equipment or drops are not adequately sterilized or restricted to a single use, infection with *Pseudomonas aeruginosa* may occur. Infection is also associated with contaminated personal contact-lens cleaning equipment. *Pseudomonas* infection may be rapidly progressive, resulting in ocular perforation and loss of vision. Bacterial conjunctivitis presents with hyperaemic, red conjunctivae and a profuse, mucopurulent discharge. Conjunctival swabs and corneal scrapings are submitted for laboratory examination. The diagnosis is confirmed by bacterial culture or nucleic acid amplification test (NAAT) including for *C. trachomatis*. Treatment is with local antibiotics, which include fusidic acid, tetracycline chloramphenicol or fluoroquinolones.

Adenovirus infection

Half of the adenovirus serotypes have been associated with ocular infection, but types 7, 3, 11, 19 and 37are most commonly associated.
• Purulent conjunctivitis occurs, with enlargement of the ipsilateral periauricular lymph node.

• Corneal involvement leads to punctate keratitis and subepithelial inflammatory infiltration, anterior uveitis and conjunctival haemorrhages.
• Treatment is symptomatic, with antibacterial agents being used if there is evidence of secondary bacterial infection. Topical steroids should be avoided. New drugs are in early phase trials.

Varicella zoster virus

The ophthalmic dermatome of the fifth cranial nerve is involved in approximately 10% of recurrent varicella zoster virus (VZV) infections (shingles). Ocular involvement, manifests in anterior uveitis, keratitis, ocular perforation or retinal involvement. Chronic disease occurs in about one-quarter of patients. The condition is very painful and pain may continue after healing of the rash (postherpetic neuralgia). Antiviral agents (e.g. aciclovir) should be used early in the infection and may prevent complications. Severe inflammation may benefit from topical steroids. A live attenuated vaccine is available to prevent primary infection.

Herpes simplex

Ocular infection with herpes simplex is the most common infectious cause of blindness in developed countries.
• Ulcerative blepharitis, corneal involvement, follicular conjunctivitis and regional lymphadenopathy occur.
• Relapses occur approximately every 4 years.

Medical Microbiology and Infection at a Glance, Fourth Edition. Stephen H. Gillespie, Kathleen B. Bamford. © 2012 John Wiley & Sons, Ltd.

- Initially, a dendritic ulcer is present, but the later clinical picture is dominated by inflammation in deeper tissues, keratitis, corneal oedema and opacity.
- Primary infection and early relapses are treated with topical aciclovir. Steroids worsen the keratitis.
- Progressive scarring that follows repeated attacks leads to corneal opacity and is a common indication for corneal grafting.

Ocular manifestations of AIDS

'Cotton wool spots' are a common retinal manifestation of untreated HIV infection due to infarction of the retinal nerve fibre layer. Cytomegalovirus (CMV) and fungal infection may develop in patients with very low CD4 counts. CMV causes a slowly progressive retinitis characterized by necrosis and may lead to blindness. The syndrome is difficult to differentiate from ocular toxoplasmosis or syphilitic retinitis. Initial treatment with parenteral ganciclovir followed by weekly maintenance therapy to prevent relapse is used. Immune-recovery uveitis can occur after the introduction of highly active antiretroviral therapy (HAART).

Trachoma

- Trachoma is a chronic keratoconjunctivitis caused by infection with *C. trachomatis*, which is now largely confined to the tropics, where poor social conditions make transmission easier and poverty reduces access to health care.
- Symptoms develop 3–10 days after infection, with lacrimation, mucopurulent discharge, conjunctival infection and follicular hypertrophy. Treatment is with oral macrolides, such as azithromycin.
- An international campaign plan to eradicate trachoma by 2020 is under way using the SAFE strategy (**S**urgery for in-turned lids, **A**ntibiotics, **F**ace washing and **E**nvironmental improvement).

Endophthalmitis

- Endophthalmitis can develop after an ocular operation, following trauma, due to the presence of a foreign body or as a complication of systemic infection.
- Early postoperative infections are commonly with *S. aureus*, *Staphylococcus epidermidis*, streptococci or Gram-negative bacilli.
- Late postoperative infections are with streptococci or *H. influenzae*.
- Post-traumatic infections are caused by *S. epidermidis*, *Bacillus* and streptococci.
- Endogenous infections secondary to bacteraemia or fungaemia are most often with *Candida*, streptococci and enteric Gram-negative bacilli.
- Rarely, endophthalmitis is caused by the nematode *Toxocara canis* (see Chapter 54).
- The diagnosis is made by vitreous aspiration or vitrectomy specimens.
- Bacterial endophthalmitis is managed by systemic antibiotics or intravitreal injection, depending on the spectrum and pharmacokinetics of the agents.

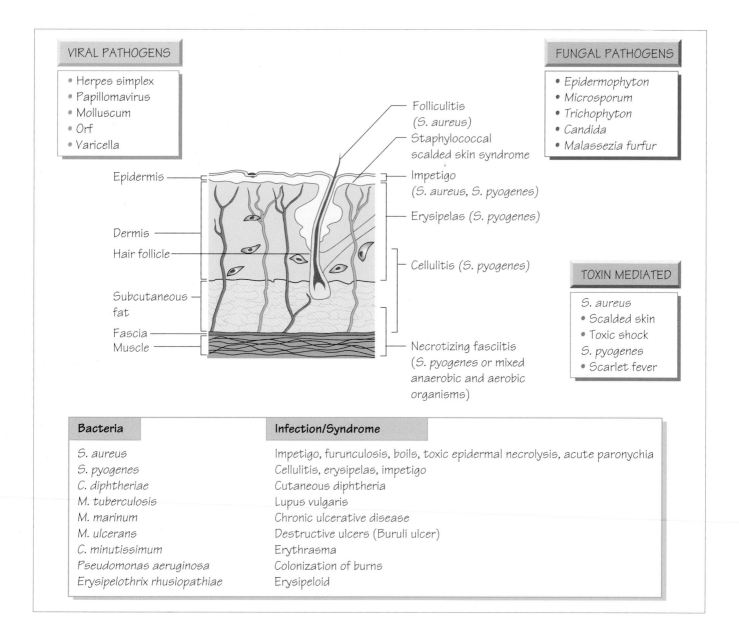

VIRAL PATHOGENS
- Herpes simplex
- Papillomavirus
- Molluscum
- Orf
- Varicella

FUNGAL PATHOGENS
- *Epidermophyton*
- *Microsporum*
- *Trichophyton*
- *Candida*
- *Malassezia furfur*

Folliculitis (S. aureus)
Staphylococcal scalded skin syndrome
Impetigo (S. aureus, S. pyogenes)
Erysipelas (S. pyogenes)
Cellulitis (S. pyogenes)
Necrotizing fasciitis (S. pyogenes or mixed anaerobic and aerobic organisms)

Epidermis
Dermis
Hair follicle
Subcutaneous fat
Fascia
Muscle

TOXIN MEDIATED

S. aureus
- Scalded skin
- Toxic shock

S. pyogenes
- Scarlet fever

Bacteria	Infection/Syndrome
S. aureus	Impetigo, furunculosis, boils, toxic epidermal necrolysis, acute paronychia
S. pyogenes	Cellulitis, erysipelas, impetigo
C. diphtheriae	Cutaneous diphtheria
M. tuberculosis	Lupus vulgaris
M. marinum	Chronic ulcerative disease
M. ulcerans	Destructive ulcers (Buruli ulcer)
C. minutissimum	Erythrasma
Pseudomonas aeruginosa	Colonization of burns
Erysipelothrix rhusiopathiae	Erysipeloid

Bacterial

Skin infections spread rapidly by contact, especially in enclosed populations or where sanitation is poor and humidity high. *Staphylococcus aureus* and *Streptococcus pyogenes* are the commonest. See Figure for others.

Disease patterns

Cellulitis
- Affects all layers of the skin.
- Causes include *S. pyogenes, S. aureus, Pasteurella multocida* or, rarely, marine vibrios or Gram-negative bacilli.
- Organisms invade via skin abrasions, insect bites or wounds.

- Empirical flucloxacillin should be given until culture results are available. Severe disease should be treated with intravenous antibiotics, including benzylpenicillin and flucloxacillin.

Necrotizing fasciitis
- A rapidly progressive infection that spreads to involve the skin and subcutaneous layers.
- It is caused by mixed aerobic and anaerobic organisms or pure *S. pyogenes*.
- The condition is characterized by pain, fever and shock, and the infected area may be discoloured.
- Progression is rapid, leading to death in a very short time.

Medical Microbiology and Infection at a Glance, Fourth Edition. Stephen H. Gillespie, Kathleen B. Bamford. © 2012 John Wiley & Sons, Ltd.
Published 2012 by John Wiley & Sons, Ltd.

- Surgical resection of infected tissue is critical, supplemented with antibiotics targeted against streptococci, staphylococci, Gram-negative bacilli and obligate anaerobes (e.g. benzylpenicillin, a third-generation cephalosporin and metronidazole).

Erythrasma

This superficial infection of the flexures is caused by *Corynebacterium minutissimum* – the lesions fluoresce under Wood's light. The organism may be cultured; treatment is with erythromycin or tetracycline.

Erysipelas

- A well-demarcated streptococcal infection confined to the dermis.
- The condition is found on the face or shins.
- It is hot and red on examination.
- A modest increase in the peripheral white blood cells and fever occur.
- Treatment is with oral amoxicillin or flucloxacillin given intravenously in severe cases.

Erysipeloid

- Zoonosis found in pig handlers and fisherman caused by *Erysipelothrix rhusiopathiae*.
- It may be self-limiting but treatment with oral penicillin or tetracycline speeds the response and is needed in rare septicaemic cases.

Burns

- Burns are susceptible to bacterial colonization.
- Typical organisms are *Pseudomonas aeruginosa*, *S. aureus*, *S. pyogenes* and, less commonly, coliforms.
- Antibiotic resistance is an increasing problem.
Complications include loss of skin grafts and secondary bacteraemia.

Paronychia

This is a common infection in community practice. The cuticle is damaged, which allows invasion with organisms such as *S. aureus*.

There is pain and swelling, followed by a small abscess. The abscess may be drained and antibiotics can be given (e.g. flucloxacillin).

Manifestations of systemic infections

The skin is a large organ that is a window onto systemic infection. Examples of this include:

- the petechial rash of meningococcal septicaemia;
- ecthyma gangrenosum of *Pseudomonas* septicaemia;
- the splinter haemorrhages of endocarditis;
- skin infarctions due to staphylococcal septicaemia;
- rash as part of a systemic infection (e.g. chickenpox and measles);
- the primary site of herpes simplex infection (see Chapter 31);
- the different skin manifestations of toxin-mediated systemic disease: toxic shock syndrome due to *Staphylococcus aureus* (generalized and palmar rash); scarlet fever due to β-haemolytic streptococci (rash with circumoral pallor, scalded skin and desquamation in neonates).

Warts

- Skin infected with human papillomavirus (HPV) shows increased replication, which gives rise to a wart. Papular, macular or mosaic variants occur. Verrucae (plantar warts) are found on the soles of the feet. The virus is transmitted by direct contact, particularly under wet conditions, such as around swimming pools.
- Genital warts (condylomata acuminata) may be transmitted sexually. Diagnosis is usually clinical.
- Virus in condylomata acuminata can be detected by immunofluorescence and nucleic acid amplification test (NAAT).
- HPV is associated with malignancy: cervical (HPV-16 and HPV-18); and laryngeal (HPV-6 and HPV-11). A vaccine has been introduced that provides sustained immunity against the serotypes associated with cervical cancer.
- Except in immunocompromised patients, warts are self-limiting and resolve spontaneously without scarring. Over-the-counter, topical keratolytic agents (e.g. salicylic acid) are widely used.
- Genital warts may respond to the application of podophyllum by trained staff. Cryotherapy is the second-line therapy

Index